Simulation with Entropy in Engineering Thermodynamics

Understanding Matter and Systems with Bondgraphs

Jean Thoma and Gianni Mocellin

Simulation with Entropy in Engineering Thermodynamics

Understanding Matter and Systems with Bondgraphs

With 87 Figures, 4 in Color

 Springer

Prof. Jean Thoma
Bellevueweg 25
CH-6300 Zug
Switzerland
www.jthoma.ch
jean@jthoma.ch

Dr. Gianni Mocellin
15, rue du Diorama
CH-1204 Geneva
Switzerland
www.straco.ch
mocellin@straco.ch

ISBN 978-3-642-06933-8   e-ISBN 978-3-540-32851-3

Cover design: Erich Kirchner, Heidelberg

For their Love and Support

Nel mezzo del cammin di nostra vita
mi ritrovai per una selva oscura,
che la diritta via era smarrita.

Dante Alighieri

Midway through our lifes journey
I found myself in shady woods,
where the direct path had faded away.

To my Wife Rose-Marie
Jean Thoma

To my Wife Anne
Gianni Mocellin

Preface

It is with great pleasure that we present this book to the public. In principle it is about thermodynamics, especially the simulation of thermofluid systems.

In popular opinion, thermodynamics is considered to be highly abstract and difficult to comprehend with its many symbols. We endeavor to show the reader how simple and beautiful thermodynamics really is.

To achieve this simplicity we apply two innovations:

For us, entropy is a substance-like concept, a kind of thermal charge, analogous to the well-known electric charge, and not the abstract and incomprehensible Clausius integral. This is by no means a new idea: apart from Sadi Carnot himself, people such as Callendar (1911), Job (1971), Falk (1976) and Fuchs (1996) all adopt the same point of view. We stress where thermal charge is analogous with electric charge and also point out the differences between them.

To represent thermal systems we use Bondgraphs (BG), which are admirably suited to this purpose. They allow us to avoid many complex equations with numerous subscripts and superscripts. Of course, literature on BG abounds, including three books by present co-author Prof. Thoma and several other books published by Springer.

We use BG more as a means to clarify the nature of physical variables and their analogies in other fields rather than from the viewpoint of electronic data processing. For example, the difference between c_v (specific heat at constant volume) and c_p (specific heat at constant pressure) is common to all multiport-Cs; and BG make this very clear.

We start chapter 1 with thermodynamics as universal science and with entropy as carrier of thermal power, commonly called heat. The difference between heat and heat flow, or of entropy and entropy flow is stressed, although they are connected by a simple integration over time. We include thermal radiation and convection by moving fluids (and electricity, chap. 1.6). We also state when simulation by entropy flow and simulation with heat flow is appropriate. The Clausius notion of entropy as a complex integral is also given, but it applies only to multiport-Cs.

Chapter 2 deals with the effects of the ever-present frictions (or irreversibilities). This includes the Carnot cycle which was invented in 1824 precisely to eliminate the effects of friction; other proponents of entropy as thermal charge are mentioned above.

In chapter 3 we consider systems with variable mass and variable mass flow. Here we use pseudo-BG with pressure and temperature as efforts, and mass flow and enthalpy flow as flows. This leads to hot gas resistors, heat exchangers and thermofluid machines.

In the following chapter 4, we apply these concepts to chemical reactions and osmosis. In fact BG can explain why some chemical reactions produce cold and not heat (the so-called endothermic reactions). This requires the concept of entropy stripping.

In chapter 5 our viewpoint changes: from considering real apparatus and machines, we descend by 20 orders of magnitude and treat only particles, atoms or degrees of freedom (DOF) of atoms, with the hope that the laws of nature are still the same. We discuss statistical aspects which apply to single DOF and relations to theory of information and its biological relevance. This chapter has been written in collaboration with Prof. Henri Atlan of Paris and Jerusalem. We thank him sincerely for his invaluable contribution. This chapter also gives some applications of entropy and information theory. It is largely based on Prof. Thoma's works during his stays at the International Institute for Applied System Analysis (IIASA) near Vienna, Austria and brings in questions of particular concern today, such as solar energy and global warming.

Appendix 1 gives tables of BG symbols which may help the uninitiated to understand this text; naturally a wide range of background reading or some familiarity with BG would be valuable.

Appendix 2 gives some notions useful for the application of BG in automatic control. There follow some historical remarks with some points that seem important to us.

Apart from the cross-disciplinary ramifications of the ideas in this book, a particular interest of Dr. Mocellin, it originates in essence from Prof. Thoma's experience as fluid power consultant and later as Bondgrapher. This involved traveling to many countries where he also arranged meetings and held discussions with people in the world of science. In Prof. Thoma's experience, nothing beats a face to face meeting for developing new ideas. Formerly we had a language problem with this, but nowadays all scientists are able to communicate in English.

The readership of this book, written more from the point of view of engineers, especially control engineers, than from that of physicists, encompasses everyone, starting with graduate students who are interested in thermodynamics and its simplicity when applied correctly, and who are also intrigued in the common structure of science across disciplines.

As this is not a textbook, there are no exercises for students, although they could be added. Indeed, part of the content has been used by Prof. Thoma for his graduate course at the University of Waterloo, Ontario, Canada, entitled: "Modern Machine Design by Bondgraphs", which included exercices for students.

Understanding the behavior of matter has always been the goal of mankind. We hope that our book makes a contribution towards that goal.

We have had discussions with many people and would like to thank them all. Most important are our wives Rosemarie Thoma and Anne Mocellin-Borgeaud, for whose patience and support during the difficult period in which the book was under construction we are extremely grateful.

<table>
<tr><td align="center">J. Thoma
www.jthoma.ch
jean@jthoma.ch</td><td align="center">G. Mocellin
www.straco.ch
mocellin@straco.ch</td></tr>
</table>

Contents

1

Thermodynamics as a Universal Science

In this book we are concerned with engineering and assume a unified and universal point of view. For this we make no distinction between ordinary heat theory and thermodynamics.

In other words, we introduce entropy immediately as thermal charge and state that the transported power equals absolute temperature multiplied by entropy current, in an analogy with electrical engineering, where the power is voltage times electric charge flow or current.

So for us, entropy is not a complex integral but simply a thermal charge; the relation to the integral is valid only in special cases as detailed in section 1.3. In his book, Fuchs (1996) adopts the same point of view and offers many applications.

We include chemistry in our book, where chemical potential is a tension-like variable and molar flow a current-like variable.

Therefore we like to call it chemical tension to bring out the analogy with electric tension or voltage. Thus entropy itself, as accumulated entropy flow, is the analog of electric charge.

Only Fuchs has dropped the first six letters from his book on thermodynamics and calls it "The Dynamics of Heat"[1].

1.1 Transmission of thermal power

A good part of our book is concerned with the transmission of entropy and heat in thermal systems, mainly because we are preparing for the simulation

---

[1] For people used to conventional dimensional analysis, we could substitute $1 \text{ J} = 1 \text{ kg s}^2 \text{ m}^{-1}$; $1 \text{ W} = 1 \text{ kg s}^2 \text{ m}^{-2}$.

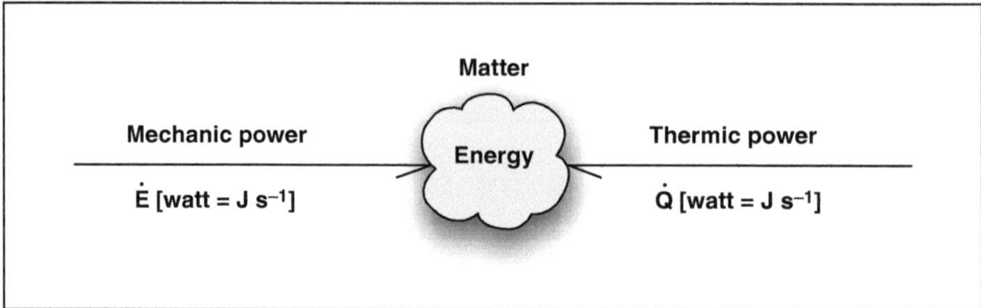

Fig. 1.1. Matter, energy and power

Fig. 1.2. Connections in BG's: bottom, true with temperature and entropy flow; top, pseudo with temperature and heat flow. Here the product of effort and flow has no physical significance.

of such systems. Here it should be observed that entropy flow and, with it, thermal power can be transmitted by three different processes:

– Thermal conduction in non-moving matter;
– Convection by moving fluids in pipes;
– Thermal radiation as in space by the sun.

These power transmission mechanisms are well known to the thermal engineer and each have different laws as we explain later. Conduction is preponderant in electricity but is only used occasionally in thermodynamics, such as through house walls (sec 1.2). More important is the convection of thermal power by a moving fluid, to which we give most of our attention (sec 3.1). Thermal radiation is important only in principle since, in practice, it is important for solar energy (sec 1.7, with remarks on the thermal balance of the earth in sec 5.5.4).

Limiting ourselves here to thermal conduction, we have simply:

$$\dot{Q} = T\dot{S} \tag{1.1}$$

This equation is fundamental and often referred to as Carnot's. It is a true Bondgraph equation where effort times flow represents a power, the thermal energy flow or heat flow; it is illustrated in Fig 1.2 at bottom. On the other hand, entropy convection in flowing fluids is qualitatively heat flow divided by absolute temperature as detailed in section 2.4. The more frequent case of heat flow or enthalpy flow in fluid pipes will be examined in section 3.1.

We have already mentioned heat conduction through walls. For this there are two kinds of BG: the true BG with temperature and entropy flow (Fig 1.2 bottom) and the pseudo-BG with temperature and heat flow (Fig 1.2 top); both are used in practice as we show in section 2.4. This is perfectly possible, but the advantages of a true BG are lost.

Accumulation equation

The accumulation of internal energy in a piece of matter, solid, liquid or gas, is given traditionally as:

$$\frac{dU}{dt} = T\,\dot{S} \tag{1.2}$$

It means that the accumulation of thermal power is equal to the (absolute) temperature times the entropy flow. This is the internal energy, conventionally called U in thermodynamics. The equation would be represented by the true BG of Fig 1.3 or Fig 1.4.

The associated energies: free energy F, enthalpy H and free enthalpy G will be described in section 1.3.

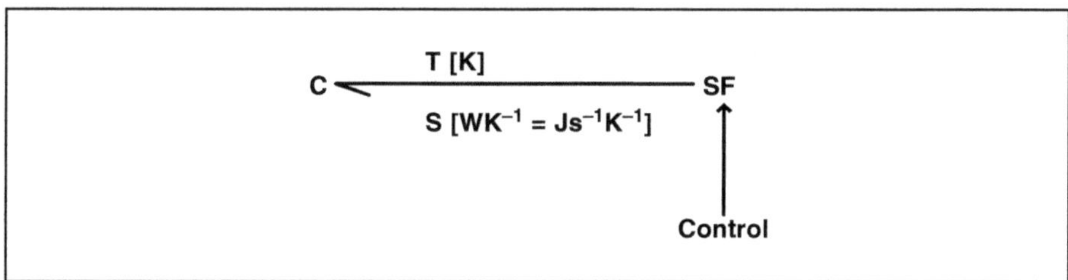

Fig. 1.3. Connection of a one-port-C and a flow source in a true BG. The vertical arrow indicates that the source can be tuned by a control system.

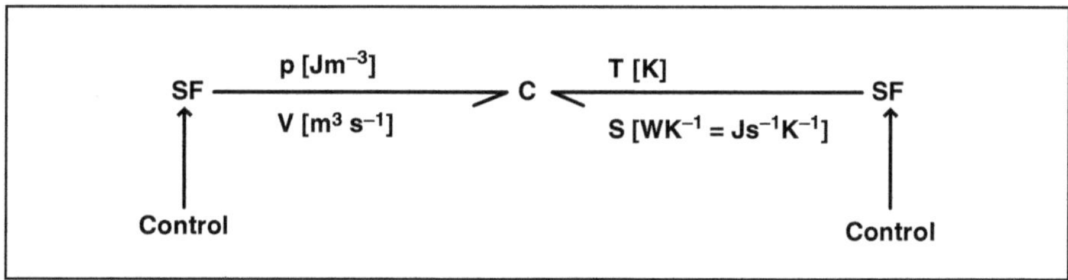

Fig. 1.4. Connections in a true BG: temperature and entropy flow at right, pressure and volume flow at left.

Note that we could integrate equation 1.2 over a time interval dt and contract it, which leads to

$$dU = T\dot{S}dt = T\,ds \tag{1.3}$$

This is the usual form in thermodynamics, but we prefer to concentrate on entropy flow and accumulation of energy as shown in equation 1.2. In other words, there are reasons for using the flows of entropy and power, not the differentials; this leads to easier physical interpretation and greater simplicity, as we shall see throughout the text.

Hence entropy is a kind of featureless thermal charge, necessarily connected with the flow of thermal power. Some people call entropy a gray paste to stress that it has no particular feature and can ooze everywhere with thermal effects; we discuss its properties later with the laws of thermodynamics (section 1.3.4). We shall also see when it is better to concentrate on entropy flow and when on heat flow (section 2.4); both exist and are needed for different problems.

To reiterate, entropy is a thermal charge, somewhat akin to an electric charge but having a quite different physical nature and properties. It is like a substance, not conserved but rather semi-conserved. This means that it can never be destroyed, but is generated by all kinds of frictions, usually called irreversibilities (Chap 2). Contrary to electric charge, entropy is not a source of an electric field and is much simpler in this respect. An interesting historical perspective, relating also to the old term "calorics", can be found in Falk 1985.

In this chapter we will concentrate on frictionless processes and elements, whilst friction with entropy generation will by dealt with in chapter 2.

In thermodynamics there can be other variables involved. For example, in a body of ideal gas, the volume can change, as we shall show later in Fig 1.8. Often the enclosed matter is a fluid such as an ideal gas. It only needs to be well stirred to have the same temperature everywhere. Alternatively, it could be a solid but then the processes must be slow enough to allow temperatures to equalize. Apart from this, thermodynamics is equally applicable to fluids and solids.

Multiport-C's are of such great importance for thermodynamics that we shall devote the next section to them, before stating the laws of thermodynamics in section 1.3.

1.2 Examples of oneport-C's and multiport-C's

Oneports

The simplest C-element has just one port and can represent a common capacitor. In the linear case, for the charge q, it has the equation

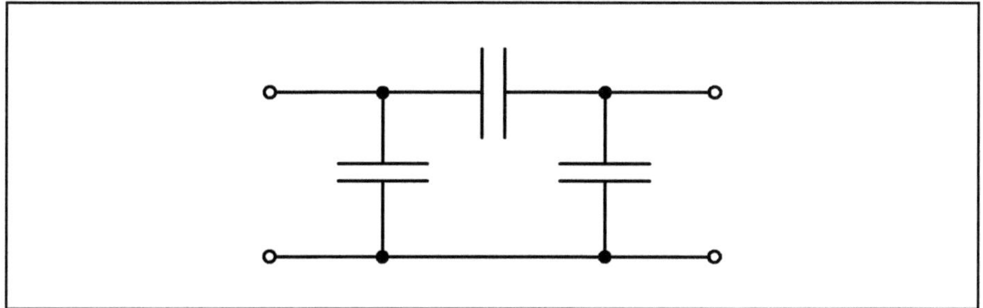

Fig. 1.5. Electrical example of a multiport-C. It is simply a capacitor network.

$$q = \int i\, dt; \qquad u = \frac{1}{C} q \tag{1.4}$$

Twoports and multiports

We can extend this to electric twoport-C's and multiport-C's. As an example of a multiport-C, Fig.1.5 shows a capacitor network of three capacitors. Here the coupling capacitor between the two capacitors on either side is important; without it the network would decompose into two separate oneport capacitors. Other capacitor networks can also be represented by multiport-C's, as long as they consist only of capacitors. In electrical circuit theory there are equivalencies between the Pi and X networks[2].

A linear multiport-C can be represented by the following equation

$$\begin{bmatrix} u_1 \\ u_2 \end{bmatrix} = \begin{bmatrix} \dfrac{1}{C_{i,j}} \end{bmatrix} * \begin{bmatrix} q_1 \\ q_2 \end{bmatrix} = \begin{bmatrix} \dfrac{1}{C_{11}} & \dfrac{1}{C_{12}} \\ \dfrac{1}{C_{21}} & \dfrac{1}{C_{21}} \end{bmatrix} * \begin{bmatrix} q_1 \\ q_2 \end{bmatrix} \tag{1.5}$$

By Maxwell reciprocity, the coupling terms C_{12} and C_{21} in the above equation are equal.

Maxwell himself discovered the reciprocity on a mechanical example, and it is still known today as Maxwell's displacement theorem in elastic structures.

Mechanical example of a table with leg fixed to the ground

One mechanical example is a table with one leg (or a T-beam) with its stem securely fixed to the ground so that it cannot move, as shown in fig 1.6. Each flange has a point of application for the forces F_1 and F_2 which produce

---

[2] If the capacitors have leakage resistance or if there is resistance of wires, the arrangement ceases to behave like a multiport-C.

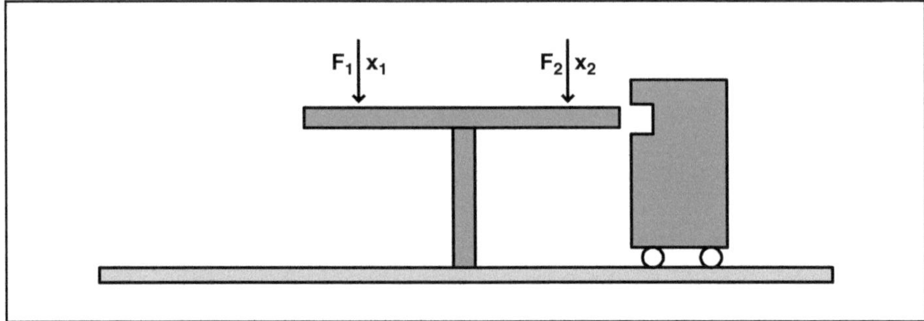

Fig. 1.6. Mechanical example of a multiport-C: a table with a central leg fixed in the ground and two points of application of forces F_1 and F_2, together with the displacements x_1 and x_2. On the right, under x_2 there is a block that can be applied and removed alternately.

the displacements x_1 and x_2. The stem and flanges have comparable elastic stiffness. There is also a rigid support which, when applied, prevents movement of one flange.

We now carry out the following experiments on the T-beam. With the support removed, we measure a certain elasticity under F_1. The flange with F_2 will move depending on the relative elasticity of the stem and flanges. If the stem is infinitely stiff, flange F_2 would not move at all.

When we apply the support under flange 2, all movement is prevented. The stiffness at flange 1 becomes higher, and flange 2 exerts a force on the support that varies with the force on flange 1. In terms of energies, with an infinitely stiff support no energy will emerge at point 2. With the support removed and replaced by a constant force, energy can emerge at this point. The T-beam is less stiff and energy emerges as a constant force with a finite value.

The whole situation is similar to that in thermodynamics: the applied support is equivalent to constant volume and the reciprocal of entropy capacity is smaller. With the support removed we have less stiffness which corresponds to constant pressure and more entropy capacity. So the difference between c_v and c_p, heat capacities at constant volume and constant pressure, is really a property of a multiport-C, entropy being the equivalent of the displacement.

An interesting and frequently asked question is: what happens if we apply so much force that the flanges go to a plastic deformation which they retain on unloading? Answer: the T-beam ceases to behave like a multiport-C.

Moving plates capacitor as an electromechanical example

Fig 1.7 shows a section through an electric capacitor where the distance between the plates can be varied. There is a force of attraction between the

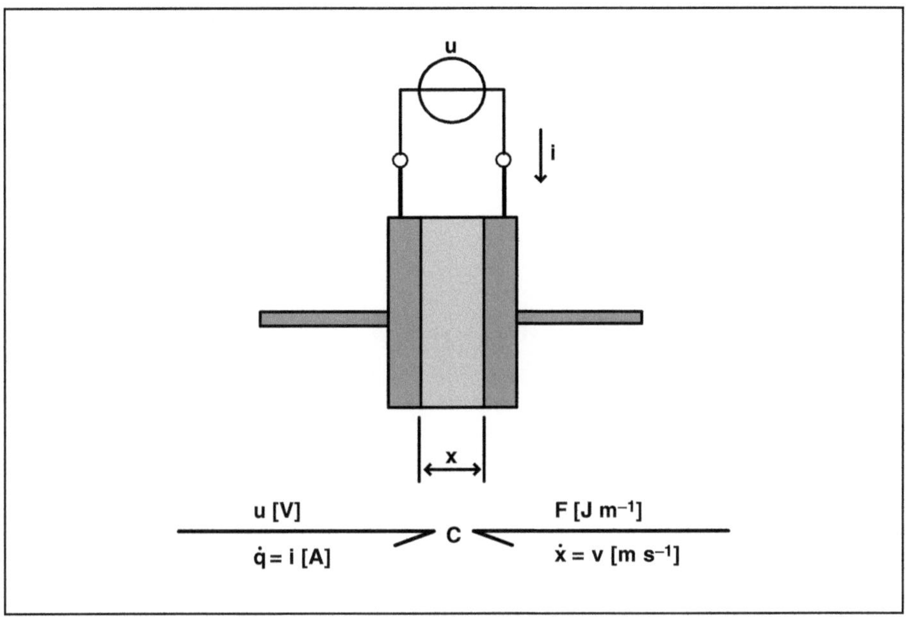

Fig. 1.7. Electromechanical example of a multiport-C. Here the electric bond has voltage and current as effort and flow, the mechanical one attraction force of the plates and displacement.

plates and it is, thus, a multiport-C with voltage and charge at the electric side and force and displacement at the mechanical side.

An interesting aspect is that this is an interdisciplinary twoport-C. Capacity C and voltage u are

$$C = \frac{\varepsilon A}{x}; \qquad q = Cu; \qquad \varepsilon = \varepsilon_0 \varepsilon_r$$

where A = area, x = plate separation and ε the induction constant. For the voltage and the attractive force we obtain

$$u = \frac{qx}{\varepsilon}; \qquad F = \frac{q^2}{2\varepsilon A} \qquad (1.6)$$

Note that here voltage and force are given correctly in terms of plate separation and charge.

The increments of voltage and force become

$$\begin{bmatrix} du \\ dF \end{bmatrix} = \begin{bmatrix} \dfrac{x}{\varepsilon A} & \dfrac{2q}{\varepsilon A} \\ \dfrac{2q}{\varepsilon A} & 0 \end{bmatrix} * \begin{bmatrix} dq \\ dx \end{bmatrix} \qquad (1.7)$$

Note that Maxwell reciprocity is also true in this nonlinear multiport-C.

One can construct for amusement an electrical Carnot cycle by using a capacitor of this type with movable plates for electromechanical energy conversion: a low electric charge is applied for separating the plates and a high charge for bringing them together. Using Carnot's invention (1824), one can interpose two phases with a constant charge and variable plate separation to adapt the voltage between the phases (see section 2.3). This is not very practical but can serve to illustrate the use of multiport-C.

1.3 Thermal multiport-C and the laws of thermodynamics

1.3.1 General

After familiarizing ourselves with multiport-C in different disciplines, most of the thermodynamics is simply contained in the properties of a twoport-C with temperature and entropy at the thermal side and pressure and volume at the hydraulic side. This is shown in Fig 1.8. So we have for the accumulation of power in a twoport

$$\frac{dU}{dt} = T\dot{S} - p\dot{V} \tag{1.8}$$

Here energy is contained in the multiport-C, which can be driven from flow sources. The multiport-C simply shows the accumulation of thermal energy through a thermal bond on the right, and the volume or fluid bond on the left. In the fluid bond, power is given as pressure time volume flow. Multiport C-elements appear frequently in physics and some examples are given in the last section. Note that there exist also multiport I-elements and R-elements used elsewhere to study other fields of physics.

The above accumulation equation 1.8 is shown in Fig 1.8 as a Bondgraph with two flow sources for thermal and fluid energy. It can be integrated (compare equation 1.3) into

$$dU = TdS - pdV; \qquad dQ = T\,dS \tag{1.9}$$

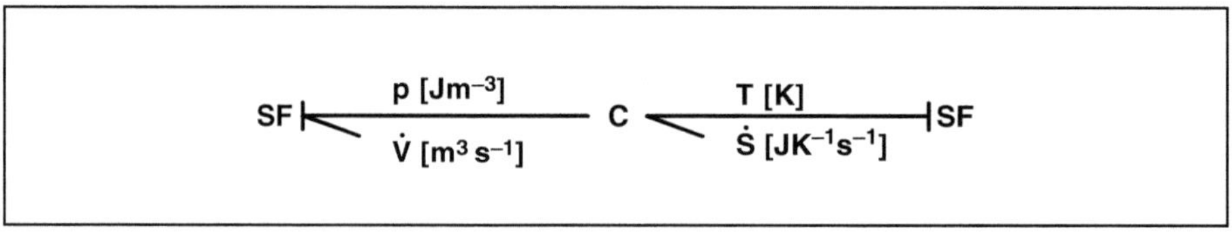

Fig. 1.8. Similar to Fig 1.2, but with two bonds and two sources in a true BG.

In many texts this is called a TdS equation. Here similar remarks apply: the power formulation in 1.8 is easier to interpret than the TdS formulation, the differentials in 2^3.

Here energy is contained in the multiport-C, which can be driven from two flow sources as shown in Fig 1.8. To continue, an accumulation of thermal energy occurs through a thermal bond on the right and the volume or fluid bond on the left.

It follows that temperature and pressure can be expressed as derivatives:

$$T = \frac{\partial U(S, V)}{\partial S}; \qquad p = -\frac{\partial U(S, V)}{\partial V} \tag{1.10}$$

Here we have S and V as independent variables. Usually a Legendre transformation is introduced at that point to obtain temperature and pressure as independent variables. This is fine but rather obscures the physical content of the argumentation, therefore we shall not do so. It is also known that the Maxwell reciprocity is determined from the cross derivatives of equation 1.10.

1.3.2 Control engineering block diagrams

Here it is helpful to use a tool from control engineering: block diagrams.

Our equation 1.8 can be represented by the block diagram of Fig 1.9. It shows the internal energy U as a function of entropy and volume in general terms.

In fig 1.10 we go on to the increments. Here the general block of Fig 1.9 splits up into two blocks with a point of addition. As stated, this is for the increments dS and dV. In the blocks, we write the gains T and V as given by equation 1.9. Therefore it can be said that Fig 1.10 is a pictorial representation of this equation.

One application of these block diagrams is to express entropy as a function of temperature and pressure, instead of temperature and volume. This is explained by Fig 1.11. Here we have one block with entropy as a function of temperature and volume, and we place before it a block with volume as a function of pressure. It must not be forgotten that volume also depends on temperature, as shown by the vertical connection on the left. We shall continue this development in the next section 1.4.

The next Fig 1.12 shows a BG with the multiport-C subject to pressure and entropy, as is usually the case.

----

[3] One author (P. Grassmann) uses an asterisk instead of the point above the extensive variable to show clearly the difference between flows (entering and leaving) and the time derivative of the strored energy in accumulation equations like 1.

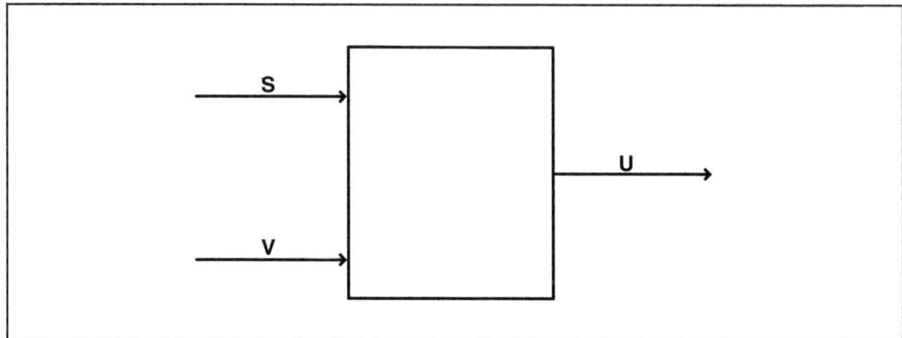

Fig. 1.9. Internal energy as a function of entropy and volume.

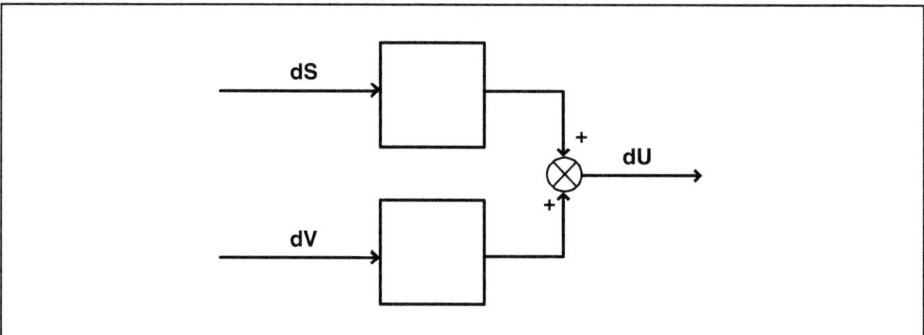

Fig. 1.10. Block diagram for the increments of the variables.

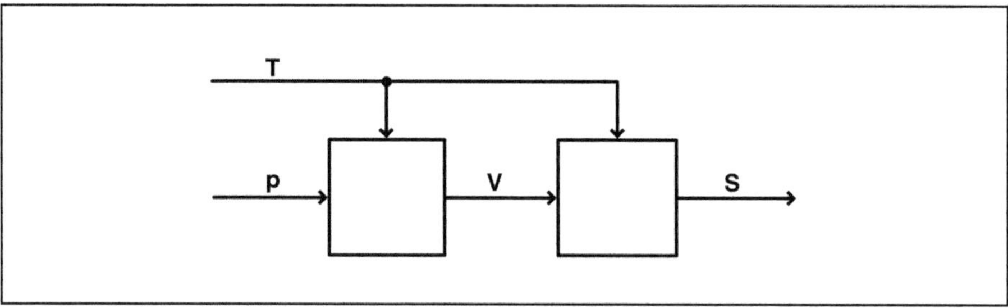

Fig. 1.11. General block diagram for entropy as a function of temperature and volume, preceded by a block giving volume as a function of pressure and temperature.

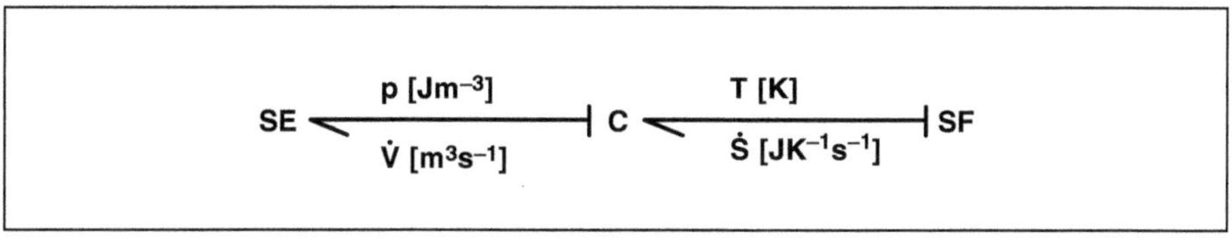

SE $\xleftarrow{\quad}$ $\dfrac{p\ [\mathrm{Jm^{-3}}]}{\dot{V}\ [\mathrm{m^3s^{-1}}]}$ $\vdash$ C $\xleftarrow{\quad}$ $\dfrac{T\ [\mathrm{K}]}{\dot{S}\ [\mathrm{JK^{-1}s^{-1}}]}$ $\vdash$ SF

Fig. 1.12. Matter under constant pressure shown as a multiport-C and a SE. Here free energy is not conserved.

In this case, on changing the setting of the entropy/temperature source, spart of the energy supplied goes into the volume/pressure source on the left. Hence the energy coming from the source on the right is not conserved. This is "elementary, my dear Watson" (quoted freely from Sherlock Homes by Arthur C. Doyle).

In many textbooks the energy emerging on the right of figure 1.12 is called free energy, but its conservation principles are never given: it is not conserved!

Box: The Bondgrapher sees immediately that Fig 1.12, with two effort sources on a multiport-C, leads to derivative causality. Hence the BG is incomputable. However, it is useful to explain the principles. If computation is required, we could apply coupling resistors, see appendix A1.

Note in particular the power orientations in Fig 1.12 and in the above equation 1.8. Power is taken positive as it flows toward the multiport-C in the thermal branch, but negative in the fluid branch of the multiport-C. This is the unfortunate convention in thermodynamics, and results in the minus signs of equations 1.8 and 1.9: this sign convention is widely used in thermodynamics, with unpleasant minus signs resulting. It would be better to make the signs all positive, but we shall adhere to convention.

Note also that matter in such a twoport-C has two DOF (degrees of freedom), S and V in this case. The associated energies mentioned at the end of this section also have two DOF.

1.3.3 Entropy after Clausius

Entropy was introduced by R. Clausius about 1875, by setting the volume constant, stating dQ = TdS and transposing

$$S = \int \frac{dQ}{T} \tag{1.11}$$

Now this equation is true at constant volume, or when, in a Multiport-C, no energy enters by or leaves from the volume bond. However, to accept it as a definition of entropy after Clausius is unnecessarily limiting.

To reiterate, entropy is a universal concept appearing in all thermal phenomena, or as already mentioned, a featureless gray paste. Sometimes, to add insult to injury, we add to dQ the suffix rev, to indicate that the heat is applied reversibly: we shall discuss its meaning in section 2.3. Other authors use a small circle around the integral sign to show that dQ is not an exact differential. For us this is a nightmare!

1.3.4 Principles or laws of thermodynamics

These principles are in all books on thermodynamics, but we shall look at them here from our own point of view. Firstly we should mention that in BG there is a distinction between power conservation and energy conservation as follows:

- Power conservation: all power is conserved in each instance and therefore energy is also conserved. Examples are junctions, transformers and gyrators.

- Energy conservation: power and thus energy can enter an element but then stays inside and somehow changes its state. The same energy must come out to return the system to rest. Hence energy is conserved in the long run. Examples are accumulators, or in general all the C and I-elements of BGs. This is true also for multiport-C and multiport-I.

Then we come to the laws of thermodynamics, called principles of thermodynamics in French and German.

First law: power and therefore energy is universally conserved; hence it is also referred to as the law of conservation.

Second law: There exists a variable entropy which is never destroyed but is generated by all kinds of friction: it is semi-conserved. Entropy is necessarily associated with heat flow, thus it is also referred to as the law of evolution.

These two laws are sufficient to study thermal effects and their interaction with fluid and other effects. However, generally one adds two more laws:

Third law: At zero absolute temperature, the entropy content of all bodies is zero (Nernst, Germany circa 1910). There is one important proviso: entropy content can be frozen in a body on rapid cooling, important for glass manufacture. In that case, the entropy content is higher also at absolute zero and entropy is calculated only from a certain datum temperature and pressure: usually a 273 K temperature and a 0.1 MPa (atmospheric) pressure.

Zero'th law: If a body A has a temperature T and the temperature of another body B is also equal to T, then the temperatures of A and B are equal. This seems obvious but it needs to be stated to define temperature properly.

We do not describe here the associated energies such as (Helmholtz) free energy F, enthalpy H and free enthalpy (Gibbs free energy) G. These are obtained by a Legendre transformation [THOMA 2000]. We would merely say that U(S,V) is energy in terms of entropy and volume, quite a different animal from F(T,V), internal energy in terms of temperature and volume. The same holds for the other associated energies H and G, except that the associated energies are taken as functions of the different variables H(S, p) and G(T, p): the Legendre transformation is often used but does not clarify the issues.

1.4 Thermodynamics in pictures

Thermodynamics is often regarded as a very abstract subject, but it can also be shown in pictures, in the form of schematics, BGs and the well known block diagrams of control engineering.

As an example, let us start from Fig 1.13 above and go on to incremental variables. The block with the two entries dT and dp, producing dV, splits into two simple blocks with an addition point. The derivatives can be written as gains into the various blocks. We emphasise that this block diagram is a consequence of our accumulation equation and independent of the nature of the enclosed body, whether an ideal gas or not.

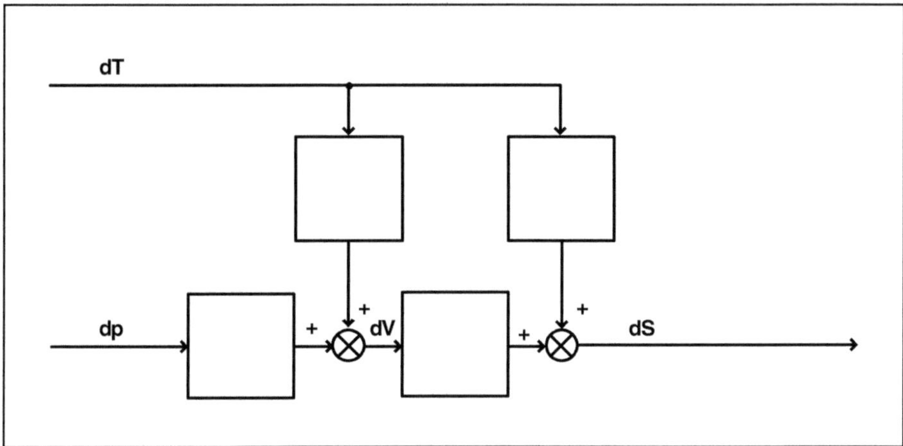

Fig. 1.13. Incremental block diagram for entropy as a function of volume and temperature, with a block for volume as a function of pressure and temperature. This last connection is easily forgotten.

As an application, we calculate the specific heat at constant volume and constant pressure. The heat flow under constant volume is

$$dQ = T\,dS = c_v\,dT$$

Which gives

$$c_v = T\frac{\partial S(V,T)}{\partial T} = TS_{-v}^{/T} \tag{1.12}$$

where c_v is the specific heat at constant volume. Note that the shortened form of the derivative is very practical and that we shall use it for our calculations, but not elsewhere. See box for explanation.

Box: Note on derivatives. In any multiport-C there appear many derivatives with certain parameters held constant. Because of this we have used the following simplified notation. This notation is very compact and can be entered in a block diagram.

We use a vertical stroke in front of the superscript to give derivation, followed by the variable to which it is applied. As a subscript we have a horizontal stroke followed by the variable that is kept constant. An example is

$$p_{-v}^{/t} = T\frac{\partial p(T,V)}{\partial T}$$

which is the increase in pressure when volume is kept constant. In the case of the rails in railway lines, this is about 3 MPa/K, which can lead to train derailment when the weather is hot. We shall use this compact notation only in the present section 1.4.

Returning to specific heat at constant pressure c_p, we support Fig 1.9 with the following mathematical development. We make entropy a function of temperature and volume, where volume itself is a function of pressure and temperature.

This was shown in Fig 1.8 in general terms and in Fig 1.9 in terms of increments dT, dS and so on. According to the rules of differentiation, the blocks with two entries split into an addition point and two simple blocks.

Now at constant volume we have dV $= 0$ and obtain equation 1.12. At constant pressure, on the other hand, we have dp $= 0$ and get equation 1.13

$$dS_{-p} = S_{-v}^{/t}\,dT + V_{-p}^{/T}S_{-T}^{/v}\,dT \tag{1.13}$$

In other words, here the volume V changes also through the vertical branch of fig 1.11 and this increases S. The next step is to invoke Maxwell reciprocity in the form

$$S_{-T}^{/v}\, dT = p_{-v}^{/T}$$

which gives, for the increases at constant pressure

$$S_{-p}^{/T} = S_{-v}^{/T} + V_{-p}^{/T} p_{-v}^{/T} \tag{1.14}$$

Now, referring to unit mass we have.

$$c_p = C_v + \frac{T}{m} V_{-p}^{/T} p_{-v}^{/T} \tag{1.15}$$

using equation 1.3 and

$$c_v = T S_{-p}^{/v} \tag{1.16}$$

Equation 1.15 is the desired relation between the specific heat values.

1.5 Case of an ideal gas

We now turn our attention to the case of an ideal gas. According to Falk, this is a general law of nature that is applicable to all matter in very low concentration, i.e. at high temperature and low density. See also Fuchs 1996, page 172,

An ideal gas has the equations

$$V(p, T) = m\frac{RT}{p}$$

$$p(V, T) = m\frac{RT}{V} \tag{1.17}$$

Inserting them in equation 1.13 we have

$$c_p = c_v + \frac{R^2 T}{pV}; \qquad c_p = c_v + R \tag{1.18}$$

The latter form is obtained by again inserting equation 1.17, and results in the well-known formula of equation 1.18.

Summarizing the procedure, we have considered that, at constant pressure, part of the thermal energy comes out in the fluid bond. Therefore the heat capacity is larger. To repeat, this variable stiffness or specific entropy, depending on what happens at the other bond, is a property of all multiport-Cs.

Many such relations can be established between the flows (displacements) and the integrals of flows of a multiport-C. In particular, there can be three or more bonds in a multiport-C, all of which have Maxwell reciprocity. The only important aspect is that frictions and irreversibilities are negligible. This is said to be obtained by running the process infinitely slowly, but we shall discuss the meaning of that statement in section 2.3.

1.6 Equilibrium in thermodynamics and electricity

To explain further the notion of equilibrium, which is often abused in thermodynamics, we illustrate it with an example: electric equilibrium. Referring to Fig 1.14, the electric voltages decrease until they become equal, whereupon the current vanishes. Note also that the voltage and current can flow in either direction, but dissipation is always positive. As a formula it can be expressed as

$$\dot{Q} = R\frac{i^2}{2} \tag{1.19}$$

This formula is valid for both positive and negative currents and always gives a positive dissipation.

See also section 3.3 for thermodynamic equilibrium. Note likewise that R must be positive.

Box: In the terminology of Falk 1980, there is a flow of carriers (extensive variable), here carriers of electric charge, and a load (intensive variable) with a load factor, here the voltage. The division between the flow of carriers and their load factor also applies, whether one uses BGs or not, to all other disciplines, such as force (load factor) and speed (flow of charge carrier) in rectilinear mechanics or torque (load factor) and rotation frequency (load carrier) in rotary mechanics. In BG terms, power is effort as load factor multiplied by flow as flow of carriers. This is one reason why BGs fit so admirably well in our system of physical variables.

Remember that the resistor produces heat as long as a current flows between the two capacitors, and the heat flow is always the product of entropy current and absolute temperature. The BG below of Fig 1.14 shows this by a RS-element, which is a BG symbol representing irreversible power. In general, the RS-field is reversible on the electrical side where the electric current can reverse, but irreversible on the thermal side. This means that dissipation must always be positive. This is nicely illustrated by a linear resistor where current can reverse but dissipation not. The limiting case is reached when no current flows or no dissipation exists. The whole concept is also a good introduction to irreversibility in nature (see chapter 2).

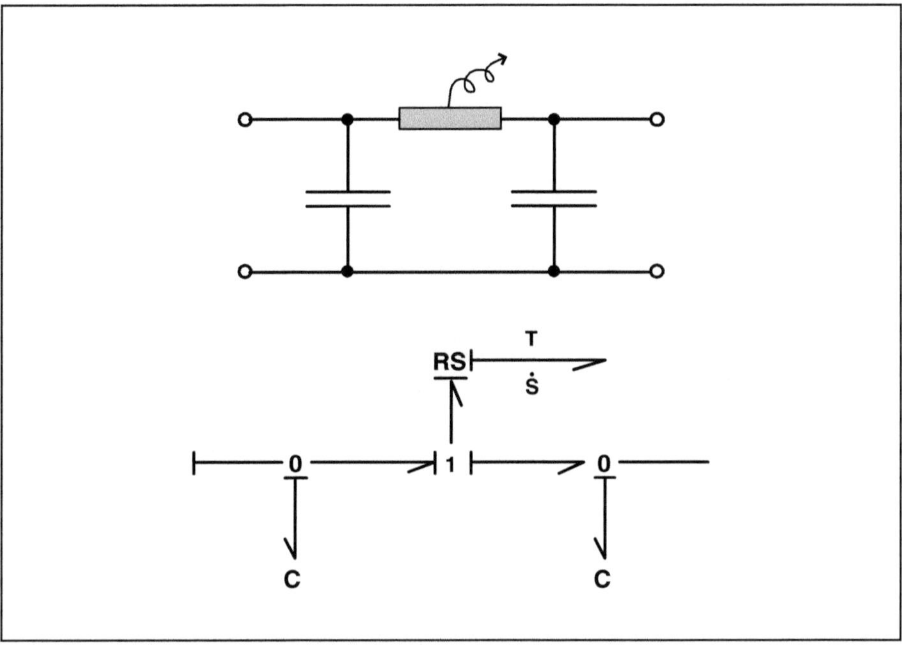

Fig. 1.14. Example of electric equilibrium with the voltages of both capacitors equal. If they are not, a current will flow through the resistor with consequent dissipation which appears as heat flow. Above circuit, below BG.

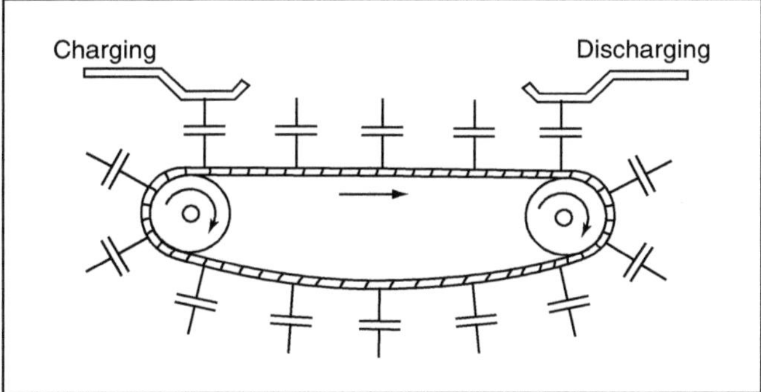

Fig. 1.15. Conveyor belt as an example of convection of electricity.

Let us mention at this point that convection of thermal energy is very important in thermodynamics, as we shall explain in section 3.1. Convection can also be conceived for electricity as shown in Fig 1.15. This is not a very practical scheme, although it bears some resemblance to the van de Graff high voltage generator, as used formerly in nuclear physics. We have a conveyer

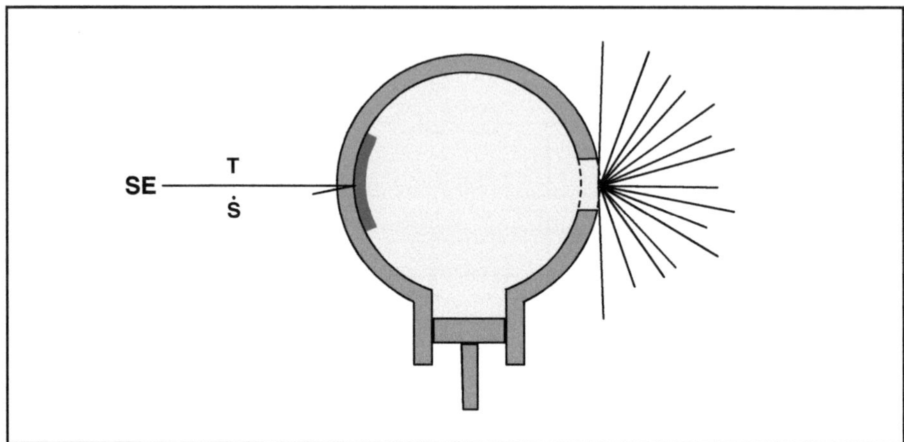

Fig. 1.16. Cavity whose volume can be varied by the piston below, and which is subject to radiation pressure.

belt with capacitors between a charging and a discharging station and also the necessary capacitor return. So the charging station fills the capacitors to a certain voltage u and takes it back at the discharging station. We define the current i as full charge multiplied by the transmission frequency. Then the transmitted power is

$$\dot{E} = \frac{1}{2}iu \tag{1.20}$$

Contrary to conduction, there arises a factor one half for power, which demonstrates that such a reduction factor exists in electricity as well. Also, by suitably controlling the sources, the conveyor line can be made reversible and sparks on the contacts avoided. This means that there is no voltage drop between the charging and discharging wires and the capacitors.

1.7 Thermal radiation

We come now to the third way of transmitting thermal energy: by radiation. The most prominent example is solar power. Anticipating appendix A1, we denote the sun as a black body radiator of 6000 K with a power density of 1 400 Watt per m^2 approximately[4].

To fix ideas, Fig 1.16 shows a black cavity with a small hole and reflecting walls. The hole is small enough not to disturb the radiation field inside, but

---

[4] The power of the sun is very large, but only 1.4 kW meets each square meter of the higher atmosphere and only about 1.0 kW comes down per square meter of the earth.

can allow radiation to pass from the inside into space, as shown at right by the rays in all directions. Also it can admit radiation from the outside, i.e. the sun.

Two special features of this cavity are:

Inside at left there is a heat conductor leading to an effort source that supplies as much entropy as necessary to keep the temperature constant.

In the lower part of the cavity there is a cylinder with a piston that can vary the volume of the cavity. Naturally the radiation exerts pressure on the piston.

The main properties of this cavity are that the radiation field inside has many frequencies, is independent of the material of the walls and is proportional to the volume V. So, with u energy density and U total energy

$$U = uV \tag{1.21}$$

Note in particular that the total energy is not conserved but increases with volume. The necessary energy will be supplied from the source at left. On the other hand, radiation density depends on temperature, as given by equation 1.24 below.

The radiation field is often called a photon gas. As is known, photons have no rest mass but a definite momentum P, as they travel in any direction in a cavity with the speed of light c (see Falk et al 1983, pages 39, 52).

Yet they do have a certain direction and their momentum is given by this direction.

$$p = \frac{U}{c} \tag{1.22}$$

The radiation field inside will exert a pressure

$$p = \frac{u}{3} \tag{1.23}$$

Note that the dimensions are correct: pressure is energy per volume, thus an energy density, as is u.

Equation 1.23 is the basis of the following development. It was first derived by Max Plank (circa 1900) from electromagnetic considerations, but for a modern derivation see Fuchs 1996, section 2.6.2.

Fig 1.17 shows a BG corresponding to the previous principle drawing.

The cavity becomes a multiport-C, which has a normal bond at left for connection with the flow source, and a bond below to allow for a change in volume. However, on the right there is a pseudo bond for the hole with temperature and heat flow.

Hence Fig 1.17 is a combined true and pseudo BG. The bond for the hole is taken as bicausal[5] rather than as a normal bond without causality, a newer

---

[5] Bicausal bonds are often used in other disciplines.

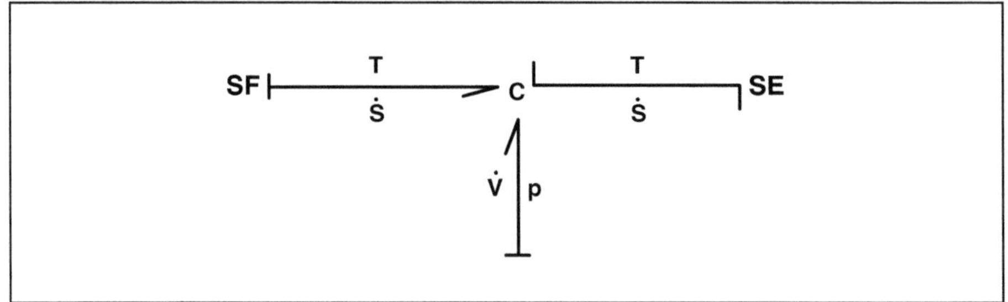

Fig. 1.17. BG of radiation cavity with variable volume. The radiation bond is described by a so-called bicausal bond.

development of BG'ing. It indicates that both temperature and heat flow come from the sun.

In fact the energy in the cavity is

$$U = uV = aVT^4 \tag{1.24}$$

with a = Stephan-Boltzman constant and u = energy density, and the entropy is (Fuchs 1996, section 2.6.3)

$$S = \frac{4}{3}aVT^3 \tag{1.25}$$

The radiation entering through the hole travels at the speed of light and disperses hemispherically as indicated by the rays in Fig 1.16. Travelling at the speed of light, it carries the above entropy and energy density along.

Hence we have, with A denoting the area of the hole, for the entropy current

$$\dot{S} = \frac{4}{3}AcaVT^3; \qquad \dot{Q} = AcaT^4 \tag{1.26}$$

where c = speed of light and for the heat power

$$\dot{Q} = \frac{3}{4}\dot{S}T; \qquad \dot{S} = \frac{4}{3}\frac{\dot{Q}}{T} \tag{1.27}$$

The first of these equations is equivalent to Carnot's equation 1.1 We see that the heat transfer in thermal radiation is only 3/4 of entropy current times temperature: not breathtaking but interesting.

If the volume of the cavity is increased in Fig 1.16 and 1.17, the source must keep the temperature constant by supplying the necessary entropy. This is the essential action of an effort source: it supplies as much flow as is necessary to keep the effort constant.

This energy is 4/3 times higher than heat flow divided by temperature. However 1/3 goes into the piston through radiation pressure, hence 1/1 remains

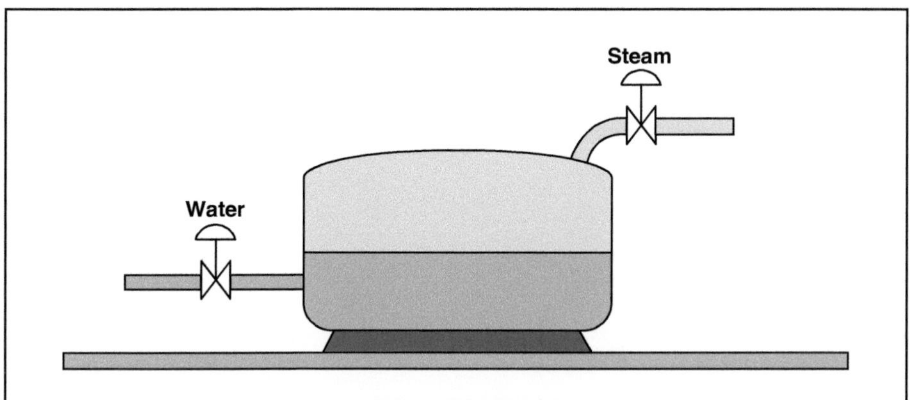

Fig. 1.18. A boiler containing some water heated from below. Here the quantity of steam is not constant but depends on evaporation and pressure.

in the new volume: entropy comes from the source but is conserved in the cavity. So the movement of the piston is entropy-conserving and reversible. All irreversibilities lie in the hole, i.e. in the transition to the pseudo BG.

Here it is very important to think clearly in terms of entropy and thermal energy (heat). Max Plank (1950) wrote in an article that he had success at the beginning of the century with cavity radiation, and later quantum theory, because he concentrated on entropy and not on heat, as his competitors did.

Also interesting is the comparison made between a radiation cavity and a steam boiler by Falk (private communication circa 1985) as shown in Fig 1.18.

The steam boiler with a small quantity of water is heated from below. There is equilibrium between water and steam at a certain pressure as given by the Clausius-Clapeyron equation.

If the heating is increased, some water evaporates, compressing the steam, until there is a new equilibrium between water and steam at a higher pressure. So the mass of steam is not constant or conserved. This is similar to what happens with radiation cavities and also with light, where quantities are not constant but depend on external parameters like temperature.

2

Frictions and Irreversibilities

2.1 Frictions of all kinds

In this chapter we consider the ever present effects of friction. Very simply, all friction takes place in R-elements, which here in our discourse become RS-elements. The S part is a source and produces the corresponding entropy, which is indestructible in our world but can be conveyed to the environment by cooling. Hence it can diminish in a machine or in a certain area if it is transported away, in practice, by fluid flow.

Returning to our resistor or R-element, alone it is a power sink, but by including thermal effects it becomes power conserving: all dissipated power becomes some new entropy (as indicated by the S part). Also worthy of note is that RS-elements are irreversible, because no entropy ever enters such elements. This is in practice the only irreversibility that exists in nature.

In addition, we compare true BGs with temperature and entropy current, and pseudo BGs with temperature and heat flow. Both have their reasons for existing, but in heat flow the pseudo BG is much simpler, as we shall see in section 2.4.

In contemplating friction, we observe the following effects:

1. Mechanical, electric and hydraulic friction;
2. Heat conduction under a finite temperature difference;
3. Chemical friction in reactions;
4. Diffusion and mixture of different substances.

Fig 2.1 gives different examples of friction and Fig 2.2 their bond graph representation.

An example of the first type of friction is the electric resistor that becomes hot and gives away heat by cooling. The next type is represented by a disk brake with a disk and brake pads, both of which must be cooled, a process

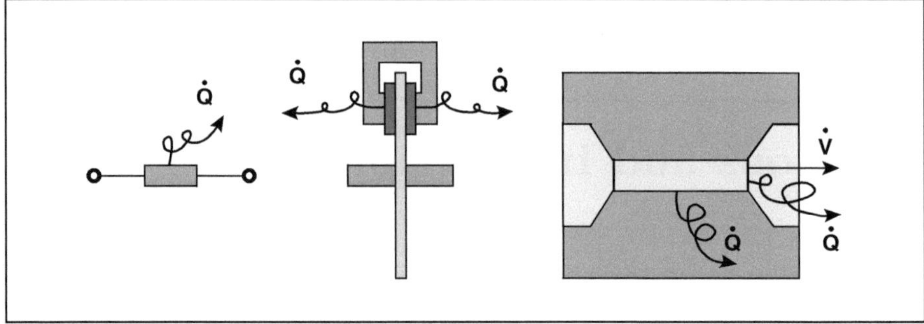

Fig. 2.1. Different types of friction: electric, mechanical, and hydraulic (the mechanical by an imaginary disk brake). All these elements produce friction heat that must be cooled away in a power conserving process.

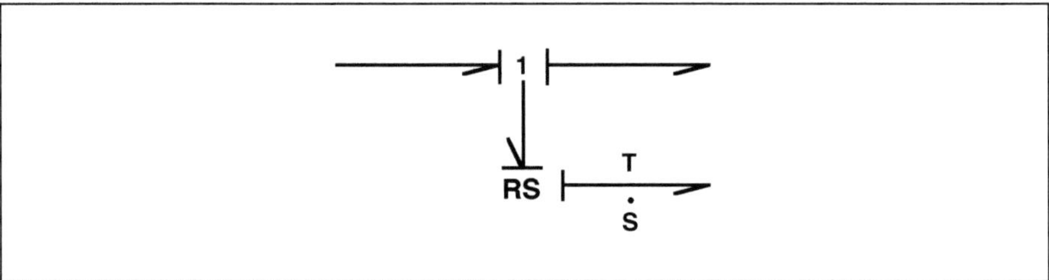

Fig. 2.2. Bond Graph of friction.

often more difficult than in the electrical example. The third type is exemplified by a hydraulic resistor in incompressible hydraulics, known as fluid power in America. It is characterized by very high pressure, up to 30 MPa approximately. Here the dissipation produces heat, both by heating the resistor (external dissipation) and by heating the flowing fluid (internal dissipation), normally oil. This heating effect of the fluid is an important feature in fluid power systems, representing about 5.5 K per 10 MPa in the outflow, whilst the external dissipation can usually be disregarded.

The next friction effect in the list is heat conduction under a finite temperature drop, to which we come in the next section, whilst item 3 (chemical reactions) will be considered in chapter 4. Diffusion (item 4) is normally considered an entropy creating process, however we will describe non-dissipative or reversible diffusion in section 4.5.

2.2 Heat conduction over a finite temperature drop, and combined conduction

Fig 2.3 is a true BG for entropy and heat conduction under a finite temperature drop. It is a resistor on a 1-junction placed between two 0-junctions, where the efforts and the temperatures, both upstream and downstream, are

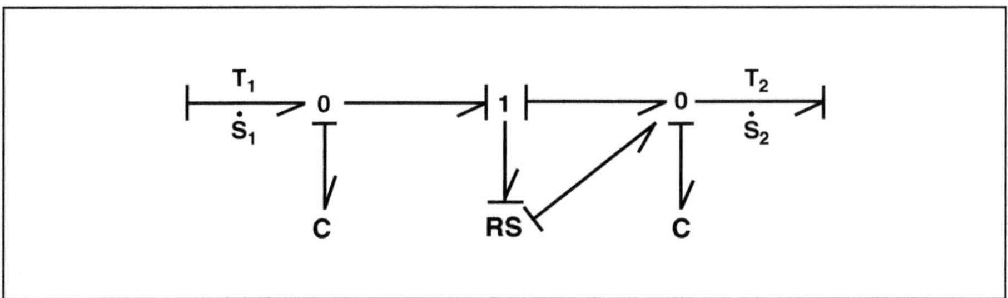

Fig. 2.3. True BG for heat conduction. Here the new entropy remains in the thermal domain and must be injected at the lower temperature[1].

calculated (set) by the C-elements nearby. Hence, the dissipated power in the RS-element in the centre is injected in the downstream 0-junction, which is at a lower temperature, as the product of temperature and new entropy flow.

The RS-element is therefore a multiport which takes the difference in temperatures and calculates the entropy flow, just like a resistor in conductance causality. In other words, it calculates the dissipated power and injects it in the 0-junction at right where the lower temperature prevails. Hence the direction of heat flow is from left (higher temperature) to right (lower temperature). So here we obtain a power conserving and irreversible multiport.

To summarize, one can also speak of dissipation as given by entropy flow and temperature difference. It produces new entropy which is injected at the lower temperature. The reticulation can be made with only one multiport RS, or it can be made with several conducting elements cut into slices, each slice being a multiport RS connected by interposed junctions and C-elements.

If the thermal members are suddenly connected in a kind of switch, the resulting arrangement can be described by Fig 2.3, except that the resistance is very small. In this case we obtain, similarly as in electricity, a thermal short circuit. This can be avoided by the Carnot cycle (next section).

If the direction of temperature drop is not certain, the dissipated entropy flow can go in either direction. In that case a SW^2 element must be placed between the multiport RS and the adjacent 0-junctions, as shown in Fig. 2.4. It has the following tasks

1. It allows the lower temperature to act on the multiport-RS;
2. It switches the entropy from the multiport-RS to the lower temperature.

Such a SW element is very easy to produce on a computer.

---

[1] The bondgrapher sees immediately that the common entropy flow comes out from the RS-element and the temperatures from the C-elements.

[2] Formely, we would have called it SWIT but SW seems so much shorter.

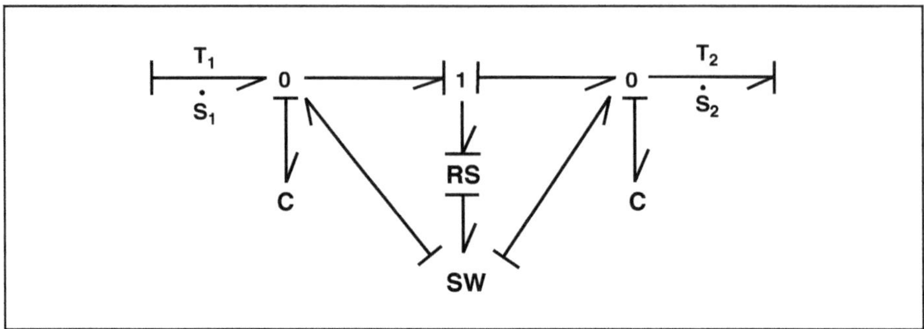

Fig. 2.4. Bondgraph similar to Fig 1 but with a switch SW added to show that the temperature drop is always responsible for entropy generation.

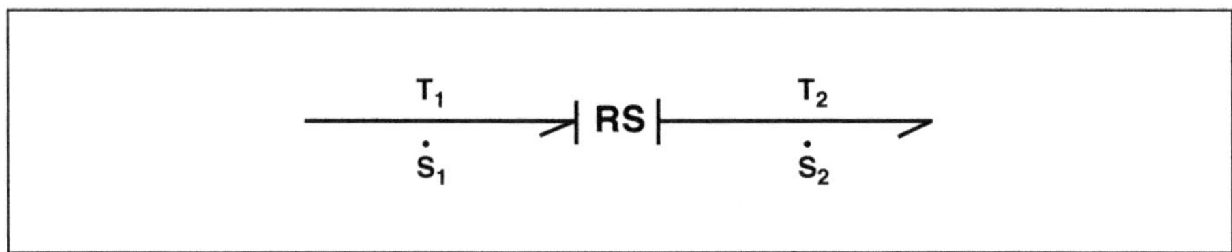

Fig. 2.5. Simplified representation of heat conduction, useful for considering principles like the Carnot cycle.

The equation for simple thermal conduction is

$$\dot{Q}_1 = \dot{Q}_2, \qquad T_2 < T_1 \qquad \dot{S}_2 > \dot{S}_1 \tag{2.1}$$

and the one for the entropy flow generated is

$$T_2 \dot{S}_{Gen} = (T_1 - T_2)S_1, \qquad \dot{S}_2 = \dot{S}_1 + \dot{S}_{Gen} \tag{2.2}$$

So the generated entropy flow is equal to the temperature drop divided by the incoming entropy flow, and, of course, divided by the downstream temperature.

In practice we include the simplified symbol in Fig 2.5 for thermal conduction. It is simply a concise representation of Fig 2.3 and will be used for the Carnot cycle in the next section.

Combined entropy and electric conduction

Sometimes we have a combined conduction of electricity and entropy, which is shown on the BG in Fig 2.6, with thermal conduction on top and electric below. It is a true Bondgraph, in which there is a multiport R with an electric

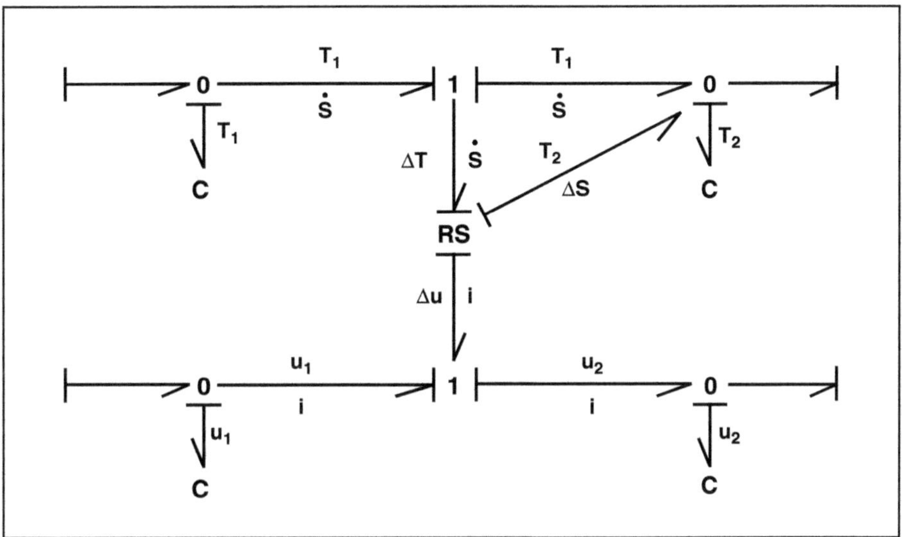

Fig. 2.6. BG for combined thermal and electric conduction, as in the Peltier effect. Here the second law requires only dissipation to remain positive, while the thermal and electric power can reverse signs.

branch (below) and a thermal branch (above). This multiport-R is, strictly, a multiport-RS. It produces entropy which is injected at the lower temperature. A prominent example is the coupling between electric current and entropy flow. In thermoelectricity, entropy flow drives the electric current, in the so-called Peltier effect, while electric current carries along the entropy flow.

Thus, combined conduction becomes the BG in Fig 2.6.

With combined conduction, the irreversibility or second law is weakened in the sense that only the dissipation is always positive. This can be expressed mathematically as

$$i(u_1 - u_2) + \dot{S}(T_1 - T_2) \geq 0 \tag{2.3}$$

This is then the irreversibility condition. In particular, the first term may become negative if the second term is larger, i.e. the total dissipation remains positive.

Some small refrigerators operate by the Peltier effect, using electricity to drive entropy from the interior to the higher temperature outside; so the inside becomes cooler. A troublesome aspect is that the Peltier elements (multiport-RS) would run in reverse if the electricity is switched off. This is alleviated by the fact that a fan cools the Peltier element, which is controlled by the same thermostat. Therefore, when both are switched off, the reverse entropy flow is at least reduced because the fan is stopped. The advantage is that it gives an inherently noise-free refrigerator (except, of course, for the noise of the fan).

2.3 Carnot cycle between two sources with friction

We come now to the Carnot cycle with two heat sources, one hot and one cold. Fig. 2.7 shows an internal combustion engine which has a certain ressemblance to the Carnot Cycle.

The device of Fig 2.8 is a multiport-C. It is driven on the mechanical side by a crank wheel (left), which is a flow source that periodically increases and decreases the gas volume. In principle the gas is connected to the hot source, as a result of which it expands and drives the crank wheel outward delivering mechanical power. At the end of the stroke it is connected to the cold source and is driven inwards, for which it needs somewhat less mechanical power. The crank wheel runs through this cycle many times over its lifetime, therefore volume increase and decrease must occur many times. This is the reason why one speaks of periodic or cyclic processes. Fig 2.8 shows the mechanism with the high temperature source T_1 and the low temperature source T_2, to which the gas or cylinder is periodically connected. These are called the isothermal phases or isotherms.

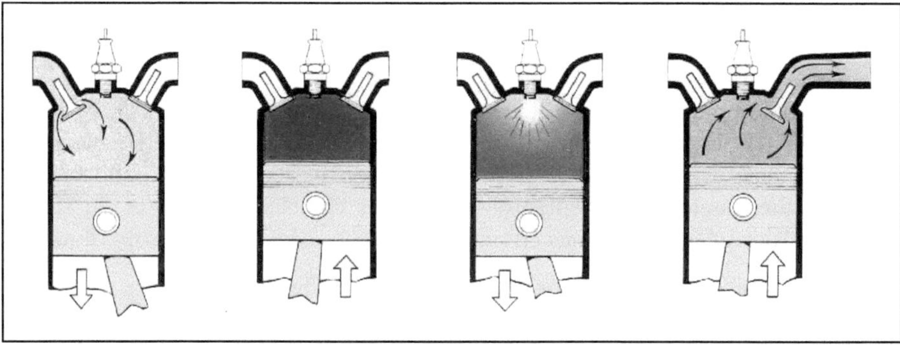

Fig. 2.7. Internal combustion engine.

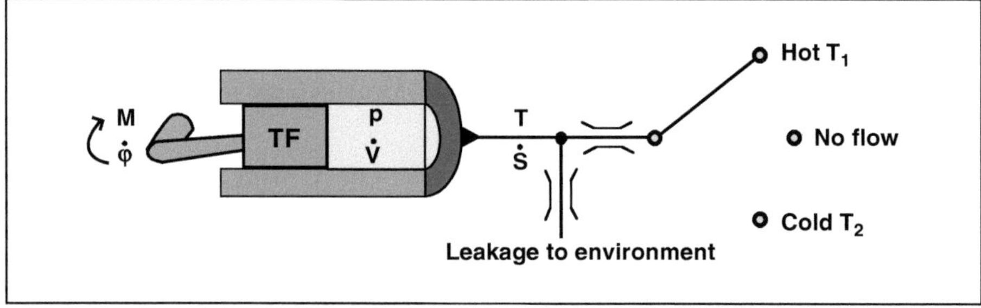

Fig. 2.8. Carnot cycle in a rudimentary heat engine as schematic. The main item is the working gas in the cylinder. The friction forces are shown by hydraulic symbols as narrow passages.

> **Box:** The Bondgrapher will note that the sources shown in Fig 2.8 impose different causalities: SE imposes an effort and SF a flow. This difficulty has spawned a copious body of literature, but to explain principles, as we do here, we need not worry about it.

Fig 2.9 shows the same mechanism as a BG with a switch or SW-element, which makes the required connections. This is not a standard BG symbol, but is very practical and replaces the three positions switch of Fig 2.8. In the intermediate position, the cylinder and the piston are not connected to any source, which is expressed traditionally by a connection to a flow source with zero flow.

With the two effort sources and the zero flow source, the causalities must change. This presents a difficulty as with electric switches, which has been the subject of much attention in the literature (see also [KMR 2000]).

It was Carnot's great invention to interpose at this point phases with zero entropy flow, i.e. with vanishing entropy flow. We show here different losses for heat conduction at finite temperature differences by multiport-RSs: heat flow and entropy gain in the conduit through RS_1 and finite heat conductivity leakage to the outside, the environment, through RS_2. A further multiport-RS would be needed for piston leakage, but this is disregarded.

The reasoning of Carnot was as follows. At the end of the high temperature part of the cycle, source SE_1 is disconnected. There he placed the intermediate phase with no entropy flow, the so-called adiabatic. Expansion continues and temperature decreases until it reaches the low temperature T_2.

Then the low temperature is connected and the entropy is pushed out. Since the temperatures are equal, no shock or thermal shortcircuit occurs. If the

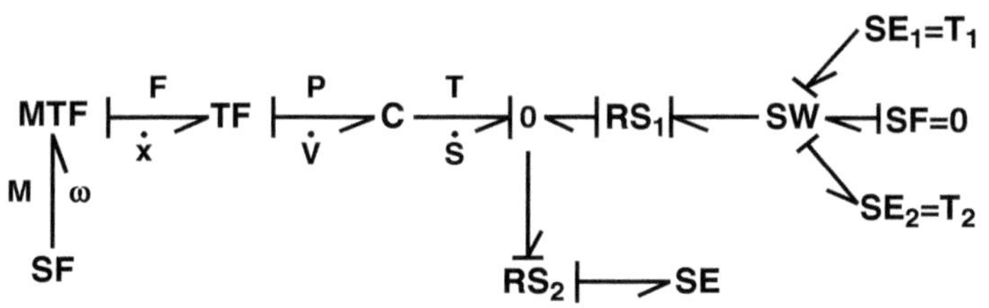

Fig. 2.9. Carnot cycle by BG similar to the preceeding figure. We therefore have several RS elements for series resistance between the sources and the working gas, and a leakage resistance to the environment.

temperatures were unequal, compensation flows would go through RS_1 and produce new entropy. These could be called thermal short circuits, which, like electric short circuits, should be avoided.

When the entropy has been pushed out, compression commences with another adiabatic or entropy-free flow, the gas compresses and the temperature rises again to T_1. This is the high temperature and the working gas is connected again to the high temperature source, whence the cycle recommences. So the temperature of the gas is adapted before reconnection to the high temperature source. This avoids entropy conduction at a high temperature difference, a kind of thermal short circuit as we said. The four phases are usually described as two adiabatics and two isothermals.

Carnot himself calls the two disconnected phases rarefaction of air without receiving caloric; we like to call it avoiding thermal short circuit, as said.

The idea of interposing two adiabatics between the two isothermals was an excellent idea of Carnot. Sometimes it is said that the Carnot cycle must be run infinitely slowly in order to avoid a temperature drop though RS_1. This is misleading because, when going very slowly, the vertical connection to the environment would produce significant entropy. Rather it must by run sufficiently slowly to make the effect of RS_1 entropy tolerably small, and fast enough to keep the effect of RS_2 entropy small as well. Whether such an intermediate speed exists depends of course on the numerical values of the various resistors or multiport-RSs.

In conventional thermodynamics, one disregards the series resistor RS_2 and RS_1 is replaced by the suffix "rev" to the supplied heat. Thoma thought about that many times in the 1960's and did not succeed in understanding it until he began using Bondgraphs in the 1970's.

A further difficulty is that one prefers to write dQ, which is conserved, over RS_1 and not dS or $\dot{S}$, which always increases. In Thoma's opinion, $\dot{S}$ is the medium that flows, not $\dot{Q}$, and it increases by flowing over any RS-element.

It is interesting to note that a Carnot cycle can also be run with a Stirling cycle that includes a regenerator [FALK 1976]. The practical difficulty of constructing the regenerator makes such economical engines in principle become voracious in practice.

So to summarize the Carnot cycle, any entropy generation in RS_1 is avoided by the insertion of the adiabatics between the isothermals, and this was his great invention. It does nothing to decrease the losses in RS_2 or the piston leakage. These are tacitly disregarded and, moreover, the whole procedure is not too clear: Thoma regards the Bondgraph of figure 1 as much more informative.

In practice, the Carnot cycle is not used because the so-called work ratio is too small, meaning that too large a part or fraction of the energy gained from expansion in each cycle must be expended to produce compression.

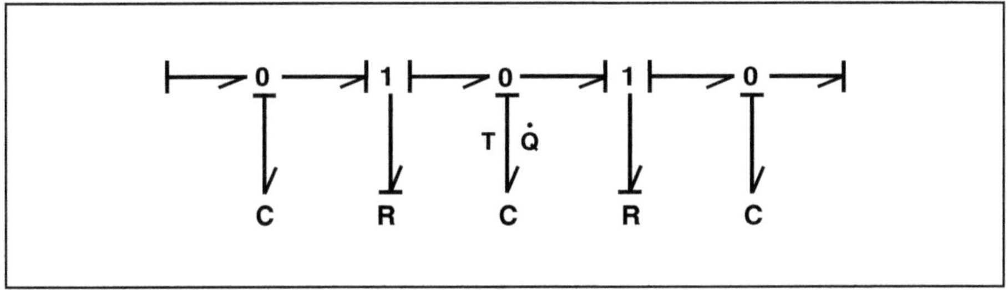

Fig. 2.10. Pseudo-BG with temperature and heat flow. Since the heat flow is conserved over the resistances, the Bondgraph is a simple R and C chain.

2.4 Heat flux and entropy flow

Excluding fluid flow, we refer here exclusively to fixed matter: it is immaterial in principle whether one considers entropy flow or heat flux; both are always connected by Carnot's equation 1.1. So in principle it makes no difference whether we take $\dot{S}$ or $\dot{Q}$, i.e. whether we write a true or pseudo-Bondgraph. The practical difference is that in conduction we have generation of new entropy (as illustrated in Fig 2.2) that must be injected at the lower temperature.

As to terminology, we can use the word "flow" when it is a flow-like variable, and the word "flux" when it is variable with the dimension of travelling power.

Hence we have entropy flow and heat flux. The difference is that entropy flow increases with thermal friction, as in Figs 2.3 and 2.4, whilst heat flow is conserved.

Heat remains constant in conduction and is conserved over the thermal resistances. So we have the much simpler Bondgraph of Fig 2.10

The pseudo-Bondgraph for heat (Fig 2.10) is very simple, a chain or R and C-elements, whilst in the true Bondgraph in Fig 2.3 we had to add the switches or SW elements, as explained.

A further practical advantage of pseudo-Bondgraphs is that we can add a power balance for the treatment of heat exchangers and turbomachines (chapter 3).

3

Mass Flows

3.1 Flow processes

In thermal power plants and chemical plants, and in thermodynamics in general, the convection of thermal power is important. It is derived from electrical power convection as dealt with in section 1.6 and Fig 1.15. However, for hot fluids we would replace the electric capacitors by something that contains fluid, say liquid bottles (incompressible) or gas bottles (compressible) on the conveyor belt. This is not very practical but shows the principle, and in practice we modify it into a fluid pipe.

A pipe for hot gas has rather similar properties to the conveyor belt. In the belt we neglected the power to drive it, although it exists. For the fluid pipe, we use a different agent to push the fluid: pressure. Hence the pressure or rather pressure drop is essential for fluid flow. The total energy flux has two parts, thermal and hydraulic.

We assume that the fluid goes slow enough for dynamic effects to be disregarded, which would not be the case for a system like in Fig 3.1 for example. In other words, dynamic pressure is always small compared to static pressure:

$$p_{Dynamic} = \frac{\rho}{2}\mathrm{v}^2 = << p_{Static}$$

Note that these velocities have to be small at the interfaces of the fluid elements only, not inside. So inside a resistor element resisting to convection, the velocities will certainly be high.

If this condition is not fulfilled and we have an appreciable dynamic pressure, we can introduce a total enthalpy flux. This is simply the static enthalpy flux plus a term consisting of dynamic pressure times volume flow rate and is written

$$\dot{H}_{Total} = \dot{H}_{Static} + \frac{\rho}{2}\mathrm{v}^2\dot{V}$$

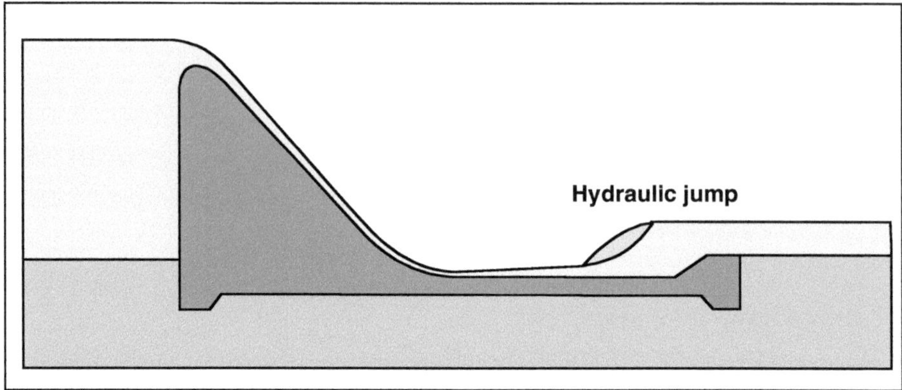

Fig. 3.1. A dam followed by a hydraulic jump. The abrupt change in depth involves a significant loss of mechanical energy through turbulent mixing.

This can be expressed as follows

$$\dot{H}_{Total} = \dot{H}_{Static} + \dot{m}\frac{1}{2}\mathrm{v}^2 \tag{3.1}$$

Equation 3.1 is remarkable for its simplicity and for the fact that it contains the flow variables of our pseudo-BG. So what remains is only the velocity v as a new DOF.

Box: The velocity v consists of two parts: radial velocity v_{rd} and axial velocity v_{ax}. The latter is not a free choice or a new DOF, but is linked to the mass flow by something such as: axial velocity is mass flow divided by an area and a mass density. So we have effectively only one new DOF, the radial velocity.

We shall not consider high velocities in this book, but they are useful for water turbines and torque converters in automobiles. An exception occurs in section 3.9 where we consider axial turbomachines with high velocity gas flow.

Returning to slow velocities at the interfaces, the total power is represented by the enthalpy flux as

$$\dot{H} = \dot{m}\,h$$

In words: enthalpy flux and mass flow are linked by the enthalpy density h (enthalpy divided by mass). It is also useful to remember how enthalpy flux and mass flow are connected in a fluid pipe, as shown in Fig 3.2.

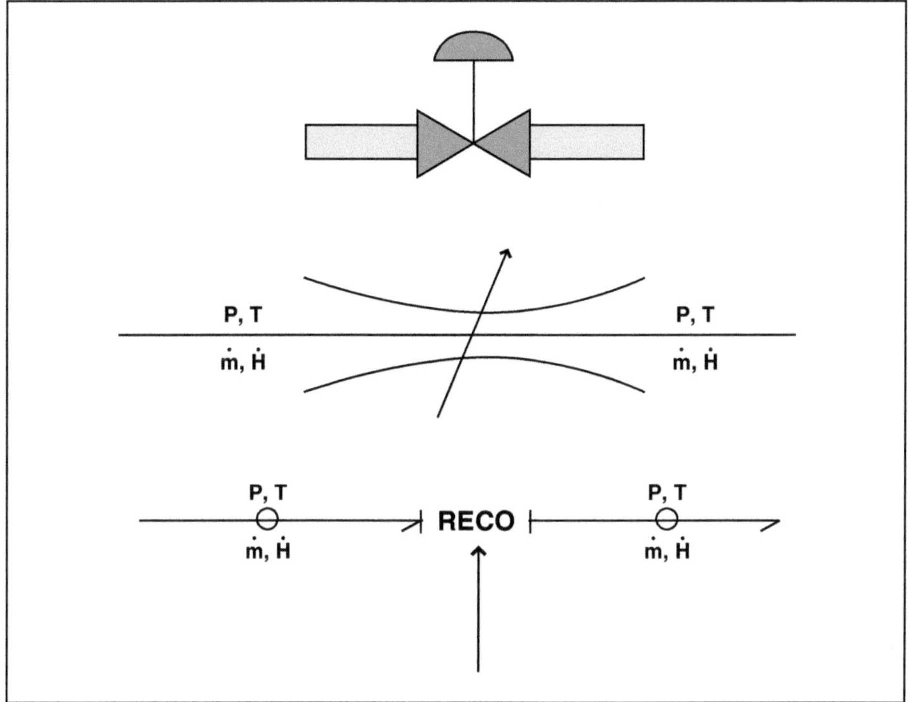

Fig. 3.2. A simple resistor shown as a narrow passage and as a pseudo BG. It conserves power, which is enthalpy flux, but produces new entropy flow.

$$\dot{H} = \dot{m}(u + p\,v) = \dot{H} + p\dot{V} \tag{3.2}$$

where u is internal energy density and v is volume density.

In so-called incompressible hydraulics or hydrostatics, one usually disregards the internal energy U and exploits only the pV part; this extends to lubrication theory and hydrostatic engineering of small gaps.

If we want to proceed to a BG representation of hot fluid pipes, it is best to use a pseudo BG, with pressure and temperature as efforts, and mass flow and enthalpy flux as flows; so it is, strictly, a pseudo vector BG. To Thoma's knowledge, this was first done by Karnopp (1979), almost 30 years ago. We have also seen representations of fluid pipes with true BGs, but they are not very practical because of the difficulty of setting values for frictions and irreversibilities.

We show a simple resistor, passage or diaphragm in center using a symbol from fluid power, and bottom, using a BG symbol, the RECO (see below near equation 3.3). This is a shortened 4-letter form of 'Resistor for Convection'. Examples include small holes, diaphragms and half open valves.

The resistor becomes a pseudo-BG, since its power flow is given by enthalpy flux alone and not by the product of efforts and flows. It is important to note that power is conserved, but not entropy flow. In the resistor there is much friction and it thus produces new entropy, normally injected in the outflow, which raises its temperature. In other words, a RECO is power conserving and entropy producing.

Box: One could be tempted to write a BG as in Fig 3.3 with a 1-junction for the mass flow.

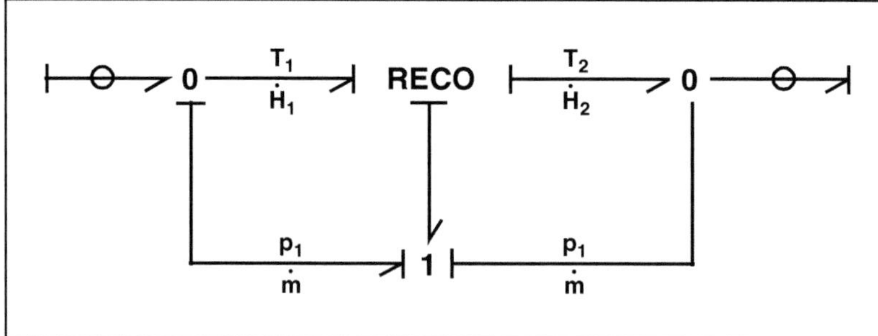

Fig. 3.3. Impossible representation of a resistor with a one junction for conserving mass flow.

This is not possible because, in a gas or compressible fluid, mass flow does not depend on the pressure differential, as Fig 3.3 would indicate, but on both p_1 and p_2 individually, in addition to the upstream temperature T_1 (see equation 3.5 below). Hence we stay with Fig 3.2; for incompressible flow, however, Fig 3.3 would be possible.

The equation for mass flow in a RECO is

$$\dot{m}_1 = \dot{m}_2 = f\left(p_1, T_1, p_2\right) \tag{3.3}$$

What is important here is that the mass flow depends on upstream and downstream pressures p_1 and p_2, and on the upstream temperature T_1, but not on the downstream temperature. In other words, downstream temperature has no effect on the mass flow through the resistor.

The enthalpy flow is coupled to the mass flow given by the enthalpy content and can be written as

$$\dot{H} = c_P \dot{m}_1 T_1$$
$$\dot{H}_2 = \dot{H}_1 \tag{3.4}$$

The second form of this equation expresses the fact that in a simple resistor the enthalpy is conserved. Mass flow is also conserved but entropy flow not[1].

Normally in gas or steam tables, one associates zero enthalpy flow with zero temperature, and counts from there. Thus T is the over-temperature, not the absolute temperature. (See next section). (See appendix A1).

A widely used form for the mass flow of compressed air in the RECO is

$$\dot{m} = \frac{K_d A\, p_1}{T_1^{0.5}} \text{ if } p_2 < 0.5\ p_1$$

$$\dot{m} = K_d A (p_1 - p_2)^{0.5} \left(\frac{p_2}{T_1} \right)^{0.5} \text{ if } p_2 > 0.5\ p_1 \tag{3.5}$$

Here T_1 is the absolute temperature. The first line applies to low counter pressure, which is also called the choked condition. The second line applies to high counter pressure, and vanishes for high values.

The functional relations in equation 3.5 look complicated with their two ranges, each described by an equation. Yet they correctly give the relation between mass flow and pressure drop in a hole, and the equations splice nicely together and even the derivatives are equal at the splicing point. Note also that at small pressure drops (p_2 near to p_1), the mass flow becomes small or zero. For negative pressure drops a SW or switch must be introduced, but we will not go into this.

More complex equations can be used if desired as, for instance, in pneumatics, but the principle remains the same.

Note also the causalities of the RECO. It requires two efforts and computes the flows: mass flow by equation 3.1 or 3.5, and enthalpy flux by equation 2, where the upstream temperature is needed.

To obtain this, the resistor must be surrounded by C-elements, or more precisely by pseudo multiport-Cs; this is shown in Fig 3.4. We call them coupling capacitors or couplers; they compute the efforts, pressure and temperature from the flows. They will be needed, for instance, if several resistors are to be cascaded.

The equations for coupling capacitors or couplers are

$$m = \int dt \left(\sum \dot{m} \right)$$

$$U = \int dt \left(\sum \dot{H} \right) \tag{3.6}$$

---

[1] From equations 3.3 and 3.4 it would follow that $T_2 = T_1$ which is false. Enthalpy of the outflow is the same. Outflow temperature decreases and out flowing entropy increases. This is not important for us because we calculate a new temperature in the following multiport-C.

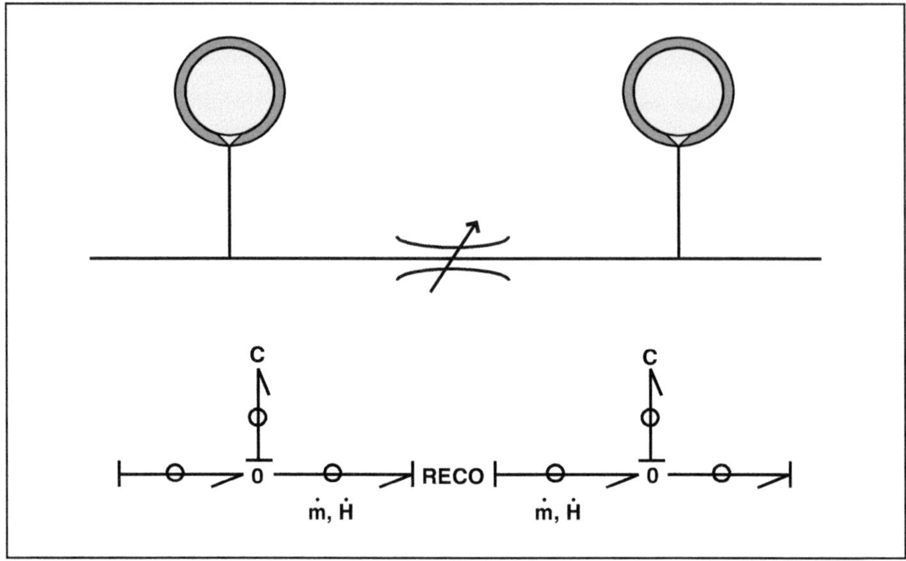

Fig. 3.4. RECO multiport surrounded by coupling capacitors, which assure correct causalities. On top as schematics with spherical capacitors, below by BG symbols. In the schematic, the resistor is adjustable by the cross arrow.

We use the letter U for the energy deposited in the coupler, because it has only stationary parts; hence it is an internal energy.

One important point is that, for the coupling capacitors or couplers, we must use the specific heat at constant volume because their walls are stationary. On the other hand, for the RECO we must use the specific heat at constant pressure c_p, because we have a hydraulic and a thermal part of enthalpy. Confusion of the two frequently leads to error.

From the mass and internal energy in a coupler we determine

$$p(V, U), \; T(m, U)$$

which becomes, in the case of an ideal gas

$$m = \frac{RU}{V\,c_\nu}$$

$$T = \frac{U}{m\,c_\nu} \tag{3.7}$$

Note that V is the geometrical volume of the coupling capacitor or coupler. It is usually the volume in the tubes between the RECOs or between other components, and as such is a fixed geometrical parameter, not a variable.

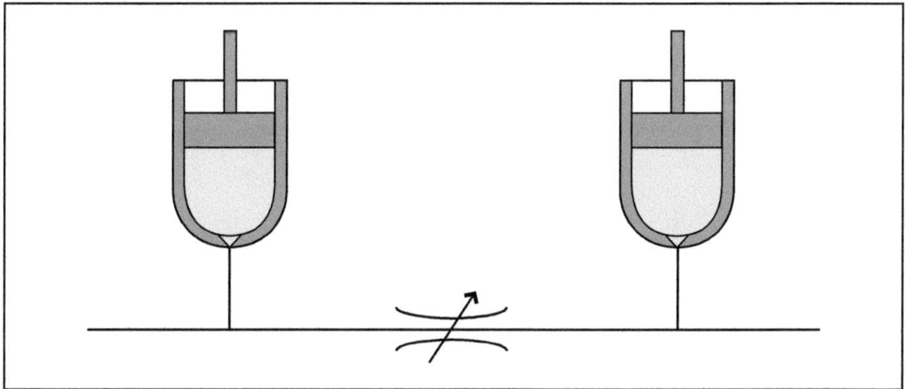

Fig. 3.5. Schematic with coupling by constant pressure, where c_p and not c_v has to be used.

If we have, as shown in Fig 3.5, pressure maintained constant by pistons loaded with constant force instead of coupling capacitors, the c_p and not the c_v has to be used.

> **Box:** One minor point of note is that the RECO produces in principle its own temperature, which could be calculated from the enthalpy and the pressures. Instead, we compute it in the next coupling capacitor, which will presumably give only negligible error.

3.2 Enthalpy and entropy in fluids pipes

For practical calculation we note that each fluid line or pipe carries a certain entropy and enthalpy, the latter being used for simulation. Entropy and enthalpy are part of the convection process and are carried by the flowing mass or mass flow.

Here we have the difficulty, as with absolute temperature, that at low temperatures, i.e. at extreme low temperatures, entropy and enthalpy are not well known. Therefore we take only over-temperatures and count all values from there, from a reference entropy and enthalpy. Usually theses references are taken under standard conditions (s = entropy mass ratio, h = enthalpy mass ratio) of atmospheric or laboratory pressure and temperature:

$$\dot{S}_{Ref} = \dot{m}s(298K,\ 100kPa)$$

$$\dot{H}_{Ref} = \dot{m}h(298K,\ 100kPa)$$

Then, actual entropy and enthalpy become

$$\dot{H} - \dot{H}_{Ref} = \dot{m}c_p(T - T_{Ref}) \qquad (3.8)$$

$$\dot{S} - \dot{S}_{Ref} = \dot{m}c_p \int \frac{dT}{T} = \dot{m}_{c_p} \ln\left(\frac{T}{T_{Ref}}\right) \qquad (3.9)$$

Here we assume that c_p is constant, which is the usual approximation. In principle there are some anomalies of the specific entropy whenever the material undergoes a phase change. This is not important for vapor, but is important for some chemical substances and solids; we will not go into this, but see our boiler example in Sec 3.6.

In other words, only over-entropy and over-enthalpy are certain and used in our calculations.

Also, it is worthwhile to note that flowing fluid has three DOF (degrees of freedom), the two DOF that all thermodynamic bodies or fluids have, and a third one represented by the velocity of the fluid, given as mass flow.

It is interesting to calculate the relation between transported entropy and enthalpy in a pipe. They are related by something with the dimension of temperature, say a calculating temperature T_{HS} as follows

$$T_{HS}\left(\dot{S} - \dot{S}_{Ref}\right) = \dot{H} - \dot{H}_{Ref} \qquad (3.10)$$

where we use the above-mentioned reference values of S and H.

Next we introduce the dimensionless parameter as follows

$$\delta = \frac{T - T_{Ref}}{T_{Ref}} = \frac{T}{T_{Ref}} - 1 \qquad (3.11)$$

This variable is then the over-temperature, over it's standard, divided by the reference temperature, a kind of nondimensional over-temperature. It allows us to express the calculating temperature as follows

$$T_{HS} = T\frac{\delta}{(1 + \delta)\ln(1 + \delta)} \qquad (3.12)$$

The function $f(\delta)$ is equal to 1 for small δ and reaches a value of 0.72 at $\delta = 1$. In practice this means that for small over-temperatures, the calculating temperature T_{HS} is equal to the actual temperature T.

Even with $\delta = 1$, that is T = 2 T_{Ref}, the function is still 72 percent; which means T = 596 K or 323 C, already a high temperature.

So up to this temperature, the calculating temperature equals the real temperature, or expressed in enthalpy and entropy, the over-enthalpy is approximately equal to the over-entropy times the (absolute) temperature. Hence to this approximation, behavior in a fluid pipe is like thermal conduction.

To resume, we have calculated a special temperature from the entropy and enthalpy content in a fluid pipe. It turns out that the calculating temperature is almost equal to the real temperature, measured in Celsius, over freezing until quite high temperatures, about 399 C. In other words: for low temperatures, the calculating temperature equals the real temperature.

3.3 Heat exchangers

The great merit of our pseudo-BG is that the treatment can be readily extended to heat exchangers and thermal turbo machines, which work by a power balance.

Fig 3.6 shows a schematic of a heat exchanger, with one central fluid tube and, symbolically, a heating shroud with a heat line passing through it, having an inflow on top and outflow below.

Fig 3.7 is the corresponding BG, or rather pseudo-BG, with one through-line fluid bond and one (pseudo) bond for the heat flow. It is called a HEXA, a contraction of Heat Exchanger. The fluid ports calculate, as before with a RECO, the mass flow from both pressure values and the upstream temperature.

Instead of the equality of enthalpy flux, we can now determine the outgoing enthalpy by

$$\dot{H}_2 = \dot{H}_1 + \dot{Q} \tag{3.13}$$

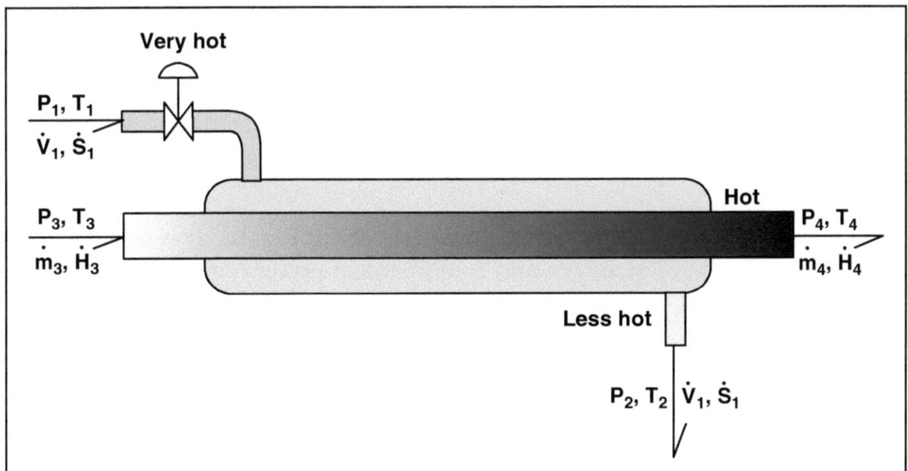

Fig. 3.6. Schematic representation of a heat exchanger. The fluid passes through horizontally.

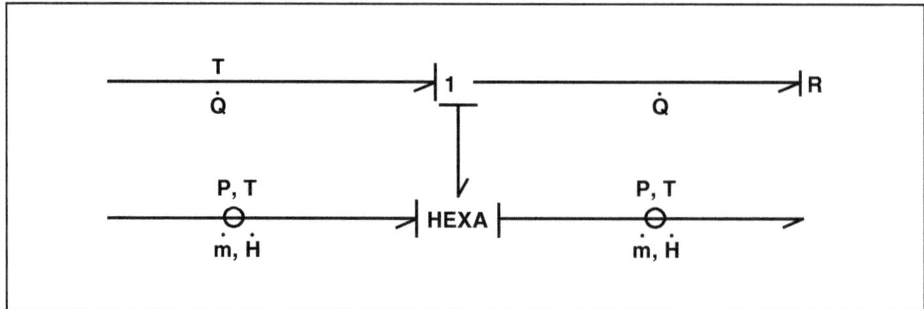

Fig. 3.7. BG of the heat exchanger. It is a pseudo for the heat flow and R calculates the heat flow.

This flux equation is realized in the HEXA. The thermal R-element is inside the HEXA and serves to obtain correct causality, that is to get heat flow from it according to equation 3.13.

Causality is determined by the fact that heat flux goes into the HEXA and temperature comes out.

The resistor at right computes the heat flux and provides the connection to the environment at constant temperature.

To summarize the heat exchanger HEXA, its equation is obtained from the resistor with its relation between pressure drop and mass flow and by adding the injected (or withdrawn) heat. This leads to the BG of Fig 3.7.

The resistor R in the heat exchanger leads to correct causality from an external temperature to the heat exchanger and gives the heat resistance of the HEXA.

Remember that we have here a pseudo BG with temperature and heat flow.

3.4 Thermal turbo machines

A thermal turbomachine is shown schematically in Fig 3.8. It is similar to the HEXA, but has several stages and a mechanical power output on the right.

This is a true bond where power equals torque times rotation frequency. It is called TEFMA, contracted from Thermal Fluid Turbo Machine. The equation is again obtained by a power balance.

$$\dot{H}_2 = \dot{H}_1 + \dot{Q} + \dot{E}_{Mec} \; ; \; \dot{E}_{Mec} = M\omega \tag{3.14}$$

The corresponding BG appears in Fig 3.9. A new feature is the mechanical branch at right, a true bond, with torque and rotation frequency. There is also an I-element which can be the inertia of the rotor.

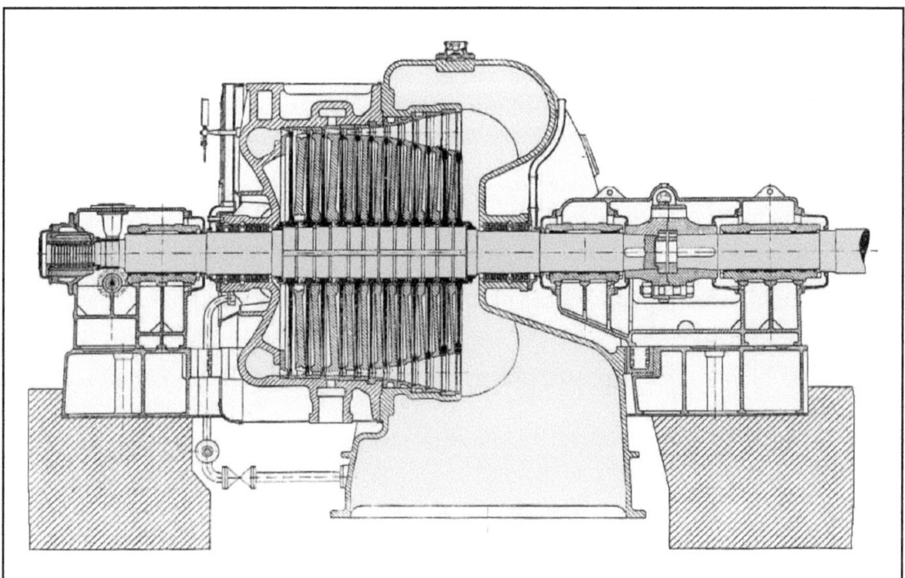

Fig. 3.8. Schematic of a thermal turbomachine.

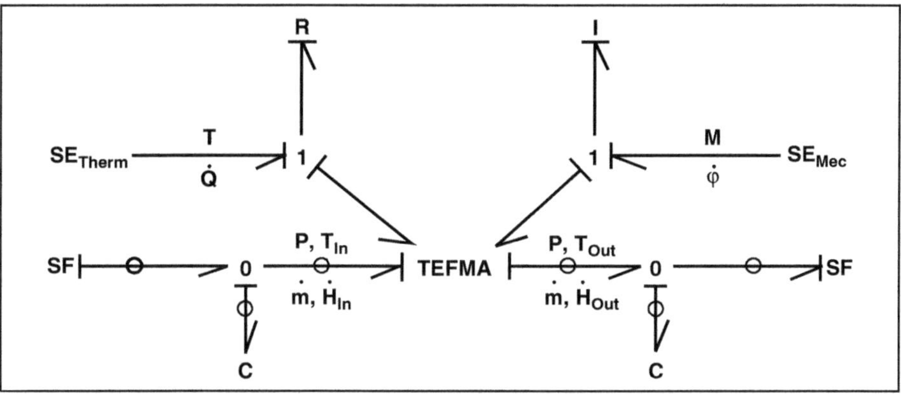

Fig. 3.9. BG for the thermal turbomachine or TEFMA of the preceding figure with mechanical bond added. Rotor inertia has been added for causality and is important in most turbomachines.

So this I element computes the rotation frequency from the torques, that is the external torque and the TEFMA torque.

For the latter we must give torque as a function of rotation frequency, the usual causality in turbo machines.

Fig 3.8 and 3.9 give the general representation of thermofluid machines. It is developed from our RECO or resistor for hot gases with the mass flow driven by the difference of pressures according to equation 3.5.

It is supplemented by the power equation 3.14.

The thermal or heat power comes at left from an exterior temperature source, shown by a SE_{Therm} over an R-element which produces the heat flow. This is then a partly pseudo-BG.

The other part of the BG is true: the rotary power at right comes as torque in function of the frequency of the shaft of the turbine and goes as output source in an SE which shows the useful torque of the machine.

An I-element, which represents the inertia of the rotor and shaft, assures the correct causalities.

Hence, we have a nice representation of a turbine, which is sufficient for most applications.

3.5 Gas flow between two vessels with scavenging

The following two-vessel application stems from pneumatic or compressed air automobiles and the process of refueling them. A car with low pressure in its tank is driven to a service station which has a larger, high pressure tank. A connection is made over a controlled resistor to fill the car tank, say from 5 MPa to 20 MPa, after what the car can be driven away again.

The performance of such a system is better than that of electric automobiles, and refuelling can by done in a few minutes, not several hours as in the electric case. Naturally, the performance is not as good as in a gasoline driven car. Unfortunately, the air in the car tank becomes hot due to compression, which reduces its capacity. According to an invention of I. Cyphelly, one adds a scavenging phase, in which a pump circulates air from the car tank to the station reservoir thus cooling the air. Since tank and reservoir are under the same high pressure, no power is needed. Of course, some friction will always be present.

Fig 3.10 shows the arrangement schematically. In the first phase refueling is begun by opining the Resistor or RECO whilst the pump is stopped.

When the pressures have equalized, the pump starts for phase two and circulates the air. The air is cooled by mixing with the air in the reservoir which continues to go into the tank over the R-element.

Fig 3.11 gives the corresponding BG with vector bonds, showing the resistor above and the scavenging pump below. Both the station reservoir and car tank become twoport-Cs, whilst the scavenging pump is a transformer in the vertical bond in the center, driven by a rotary mechanical bond. All air bonds are vectors with pressure and temperature as efforts and mass and molar flow, as indicated by the small circle around each bond.

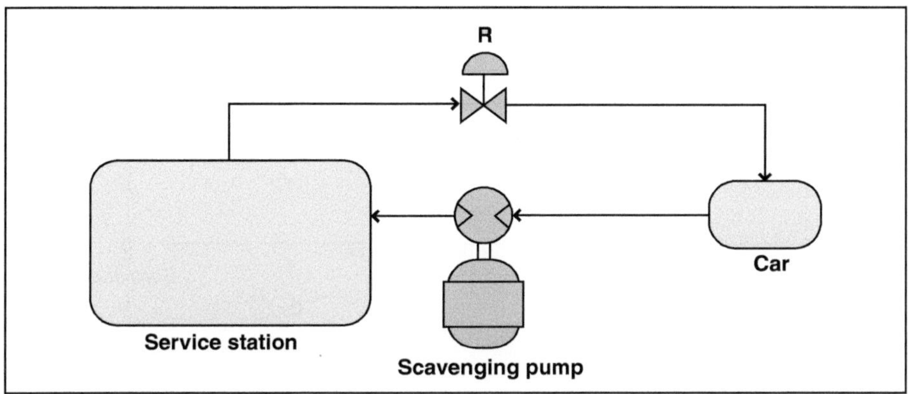

Fig. 3.10. Arrangement of pneumatic filling station with the subsequent scavenging phase.

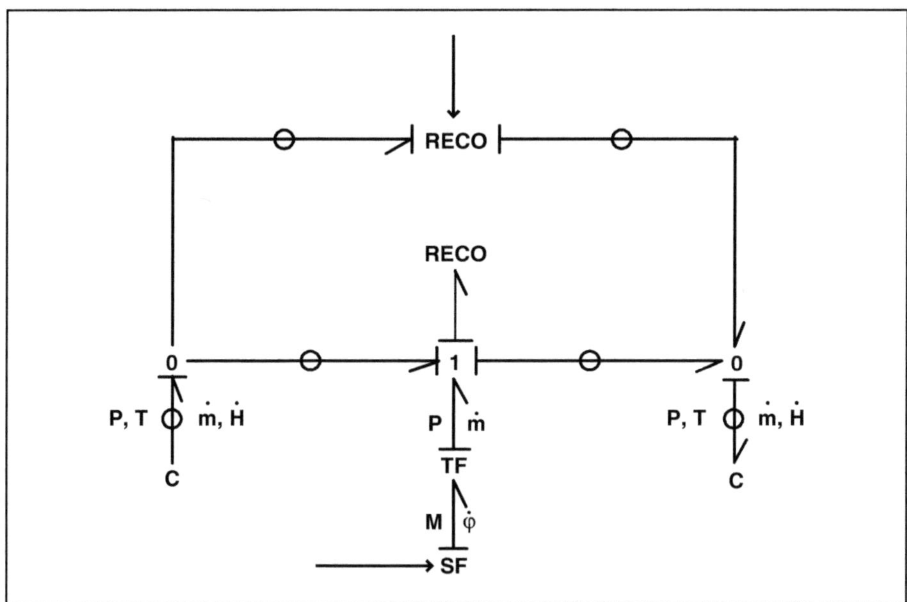

Fig. 3.11. BG for refueling with filling station and car.

Fig 3.12 is the BG for simulation written using the Twentesim program. It uses two single bonds and consequently twoport-C's for tank and reservoir. We have written two BG elements, RECO for the resistor above. The lower part gives the system for switching on the pump.

Fig 3.13 shows the final simulation result, with filling lasting from 0 to 15 s followed by the scavenging phase of 90 s. One sees that the pressure in the

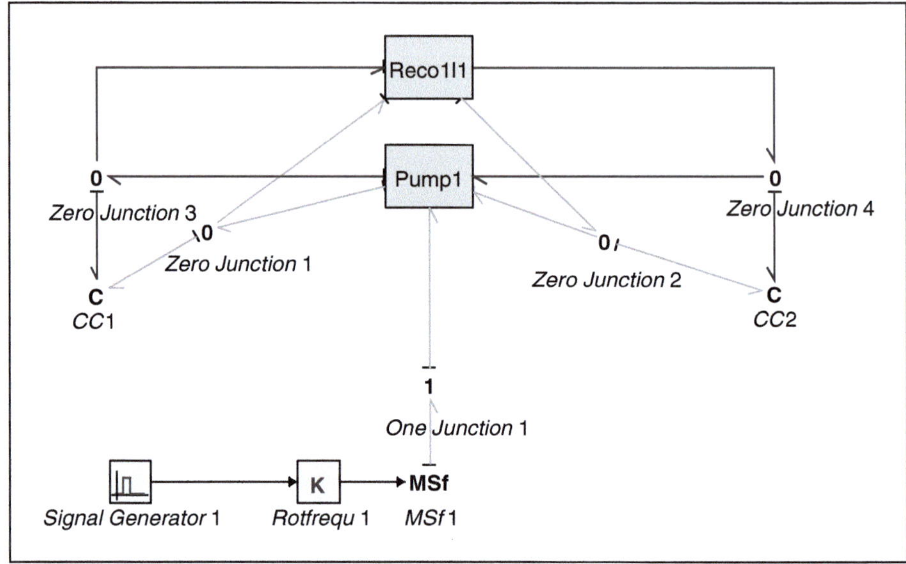

Fig. 3.12. BG written using Twentesim with single bonds and twoport-C's.

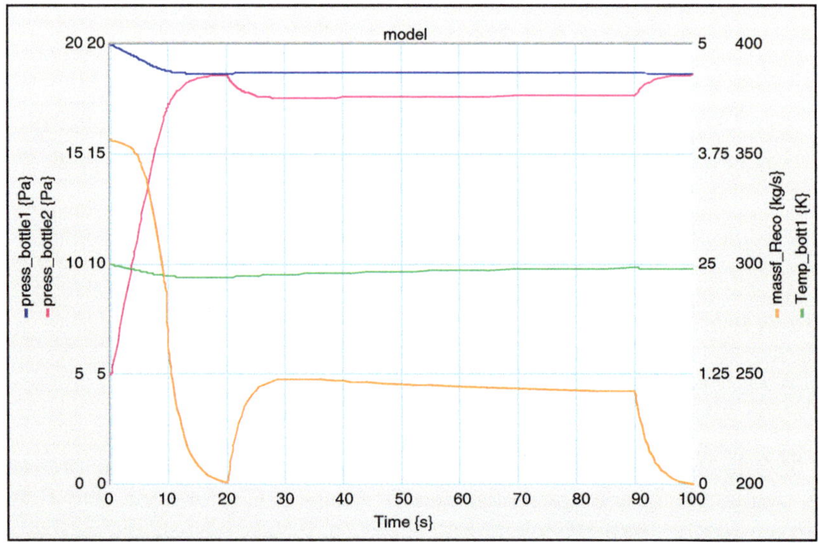

Fig. 3.13. Simulation results with a 15 s filling phase and a 75 s scavenging phase. During this latter phase, the pressure in the tank is somewhat lower.

tank falls during this phase. After 90 s the entire process is finished. This simulation was published by Thoma et al 1990 with the same parameters, and gives the same results as Twentesim does here.

3.6 Two-phase flow in boilers and condensers

One example of two-phase flow is the steam boiler shown in Fig 3.14. It is known that water-steam exchange can be considered a chemical reaction, governed by the equality of chemical tension between water and steam.

We see a boiler heated from below. Water evaporates and collects in the upper part, from which it leaves at right.

There is an automatic level control: if the water level sinks below a certain point, the valve faucet opens to deliver more water.

The next Fig 3.15 shows the corresponding BG, where the vapour phase C_{Vapour} is at the top left and the liquid phase C_{Liquid} at the bottom right.

They are connected by the mass flow, center left, and by the enthalpy flux, center right. Mass flow is controlled in the R-element by minute pressure differences between liquid and vapour phases.

The mass flow of evaporation entrains over the central MSF called Latent-Lev, the latent enthalpy of evaporation, usually called Lev

$$\dot{H}_2 = \dot{m} L_{ev} T \tag{3.15}$$

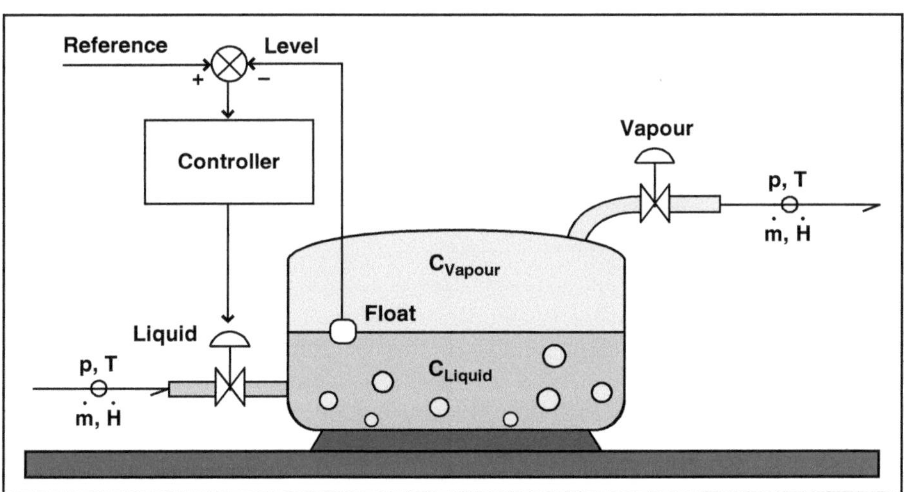

Fig. 3.14. Schematics of a steam-water boiler.

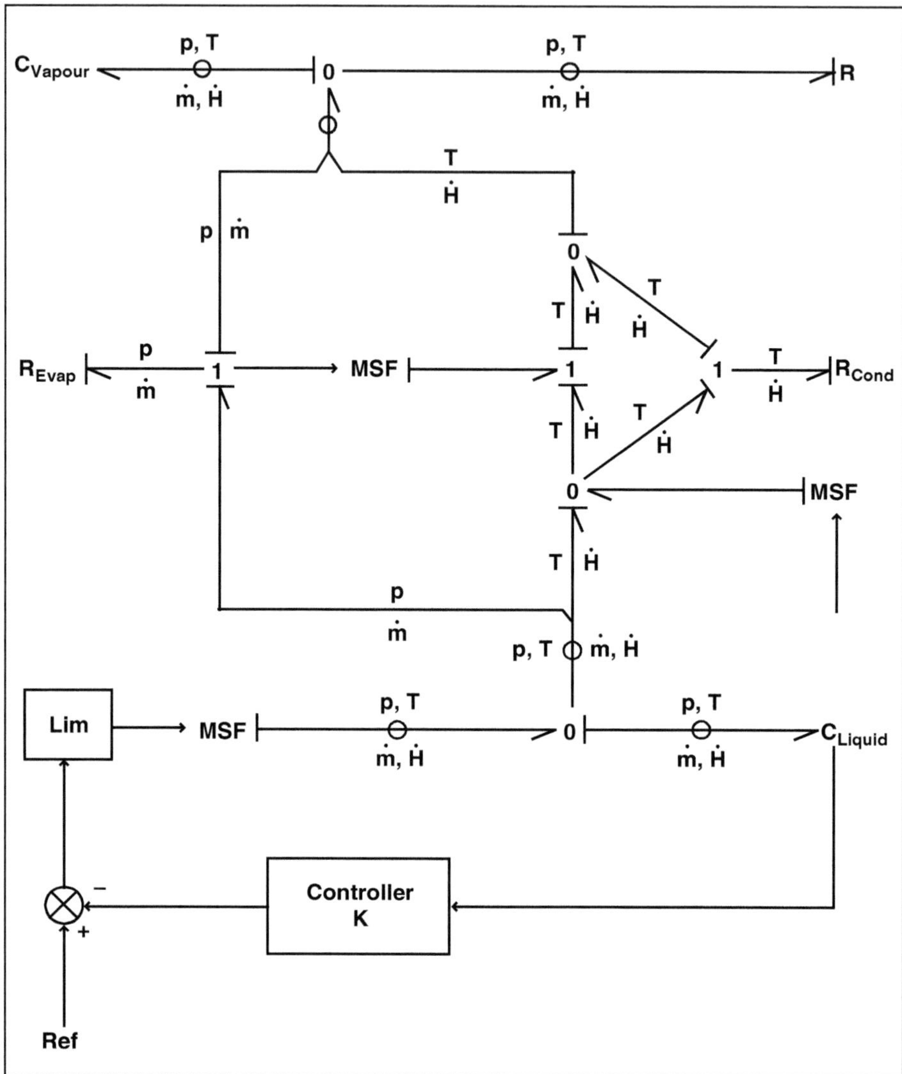

Fig. 3.15. BG for the boiler.

An important consideration is the direct heat conduction over the interface liquid-vapour by the R-element R_{Cond}. Vapour consumption is given by the R element on top, which could be modulated to simulate different steam consumption rates. Automatic liquid level control is provided by the feedback arrangement in the lower part of the figure 3.16.

R_evap is the evaporation resistance as function of pressure and temperature, with a threshold given by the Clausius-Clapeyron equation. Results of simulations are given in Figs 3.17 and 3.18.

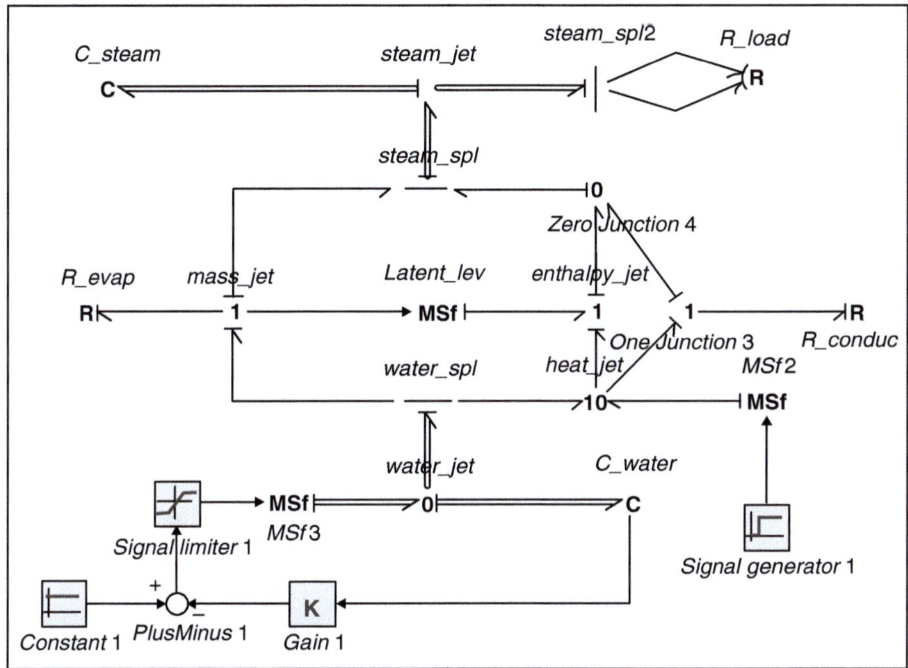

Fig. 3.16. Computer simulator of the boiler with the controller. Compared to our standards, 20-sim uses double lines instead of circles to designate vector bonds and horizontal short lines instead of our splitting triangle to indicate the transition from vector to scalar (simple) bonds.

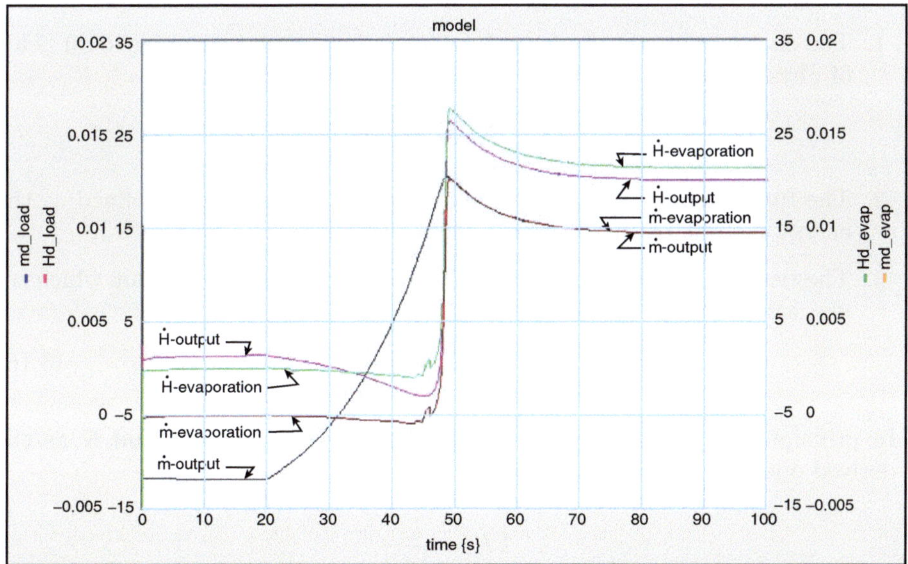

Fig. 3.17. Heating from 90 C, boiling at 100 C and supplying steam to the resistor. Right arrow: mass flow load. Left arrow: mass flow load and evaporation.

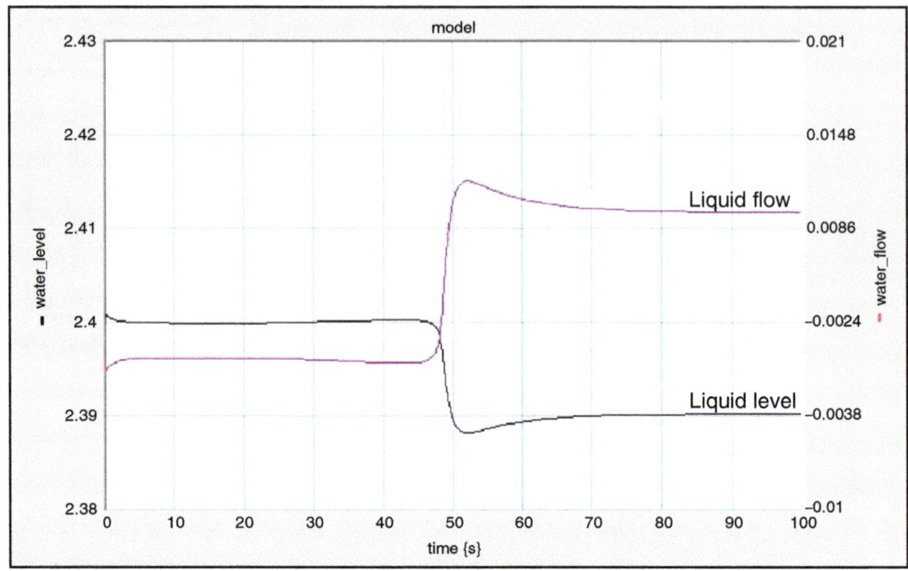

Fig. 3.18. Action of the controller: left is water flow, right the water level.

3.7 Units and overvalues in thermodynamics

The determination of units of measurement in thermodynamics and thermal engineering is by no means trivial and will be considered here. In principle we have three units, namely

1. The energy unit, the Joule or J, which is universal, belonging to all fields of physics including electrical engineering, where

$$1\,J = 1\,V\,A\,s$$

2. The fundamental temperature unit, the Kelvin, originally defined as the energy or heat required to heat 1 kg of water by 1 K, being 4.2 J.

3. The derived unit for entropy, the Carnot, not well known, but which we use, after Falk, as follows

$$1\,Ct = \frac{1\,J}{1\,K} \tag{3.16}$$

In principle one could also take Ct as the fundamental unit and K as the derived one, but this is not done.

$$1\,K = \frac{1\,J}{1\,Ct}$$

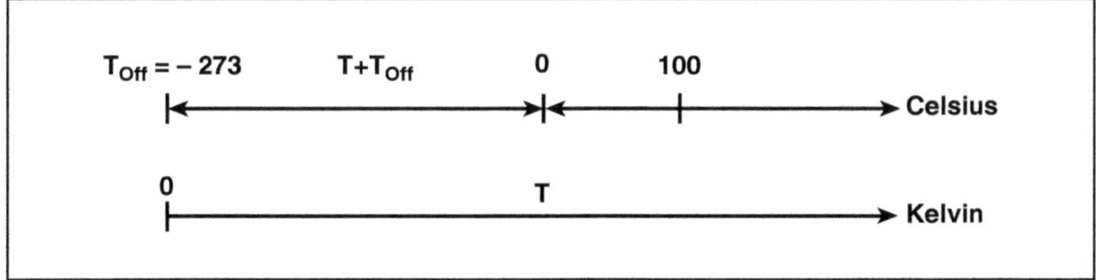

Fig. 3.19. Representation of the temperatures used in thermodynamics. The unit is always Kelvin (K), but to calculate from the melting point of ice one also uses C. T_{Off} is the offset between the two.

With temperatures, why is it that $273\,K$ is given a special unit C or Celsius and all temperatures are measured from there? It is of course the boiling point of water at a pressure of $100\,kPa$. Note that this is just a special value that still has the dimension of K. The temperature scales are shown in Fig. 3.19 and C is just the same as K, but offset by 273 K. It was determined in order to establish the melting point of ice at 0 C.

The same is true for pressure, where we have absolute pressure and overpressure. In practice for simulation we use over-temperature and over-pressure (also called gage pressure), which are $100\,kPa$ and $273\,K$. This works fine, only sometimes we are obliged to use absolute values, as in the ideal gas equation.

Note that in chemistry other origins are used, namely $100\,kPa$ and $298\,K$, i.e. 25 C laboratory temperature.

Origins of entropy and enthalpy

For entropy and enthalpy, only the difference over some datum is required. For this we take, for water and steam, 0 C temperature and $0.1\,kPa$ pressure, after Dubbel 1986. At this point entropy and enthalpy for water is taken as zero and all values are above it. For steam the entropy or heat of evaporation has to be added. This decreases with temperature and is about $2260\,kJ/kg$ for water at 100 C.

Hence our values are above these points of origin.

3.8 Exergy, an accountant's reserve

The notion of exergy, also called availability in America, appears in some thermodynamics texts and therefore we describe it here. It looks like free energy, but for the temperature one must take not the fluid temperature, rather the temperature of the environment. The convention is to take the letter

E (although more logical would be the letter U for internal energy), thus the formula is, with E_{Ex} = energy and T_{Env} = temperature of the environment,

$$E_{Ex} = E - T_{Env}S \qquad (3.17)$$

By time derivation we arrive at the notion of exergy power, that is the exergy transported through a house insulating wall or similarly through a fluid pipe.

$$\dot{E}_{Ex} = \dot{E} - T_{Env}\dot{S} \qquad (3.18)$$

So all thermal streams carry energy and entropy as expressed by equation 3.18. We can visualize exergy as an accountant's reserve for the cost of disposing of machinery at the end of their useful life. Here it is the energy minus the entropy times the temperature of the environment into which it has to be deposited. Obviously, there is no difference between exergy and energy for non-thermal domains where entropy content is zero.

The notion of exergy is very useful where there are combined thermal and electrical or mechanical considerations such as in heating systems. Running costs for such systems should be invoiced for their exergy content. The trouble is that exergy diminishes as entropy increases by thermal conduction. The worst example is an electric cooker, where electric power and energy enters without entropy; hence exergy equals energy. Thoma once proposed a district heating system based on exergy.

We conclude this section by a story: exergy expresses the fact that with much entropy content, energy has less value than in normal circumstances. Falk, about 1985, told Thoma so. But Thoma replied: it all depends on circumstances. If you are in the north of Canada dying from cold, nothing would be more precious than some entropy to heat up your body. Both agreed and had a good laugh. This was their last meeting as soon thereafter Falk died from cancer.

3.9 High velocity turbomachines

One application where high fluid velocity is needed inside a component is the turbomachine.

We start with the simple formula 3.1 to form the total power transport, that is the static and the dynamic enthalpy flux, and wish to apply it to an axial gas turbine. This machine consists of a convergent, some rotors and stators and a divergent. The idea is that in the convergent the fluid is accelerated, such that the kinetic power becomes appreciable, while in the divergent the kinetic disappears by deceleration of the gas.

Fig 3.21 is a simplified schema of a gas turbine like in Fig 3.20 with only two rotors, separated by a stator, and placed betwen a convergent and a

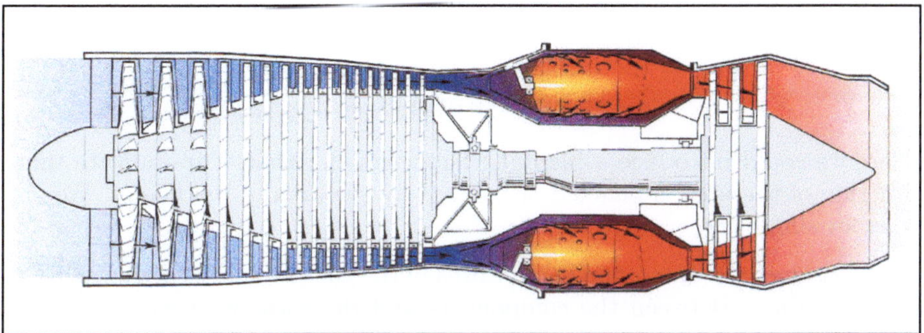

Fig. 3.20. Axial gaz turbomachine.

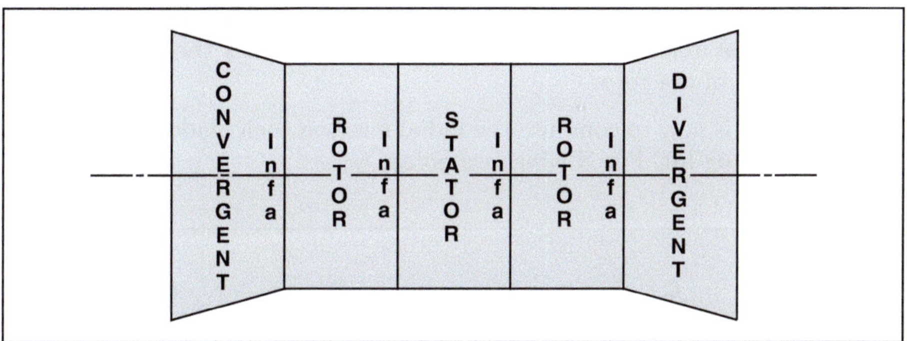

Fig. 3.21. Unwrapped parts of a gas turbine: basic arrangement of a gas turbine with two rotors, one stator, a convergent for acceleration of the gas and a divergent for deceleration.

divergent. To repeat, kinetic power becomes appreciable in the convergent, where the fluid is accelerated, while in the subsequent divergent kinetic power is absorbed. For this we have the equation

$$\dot{m}\frac{\mathrm{v}_{ab}^2}{2} = \dot{H}_{Total} - \dot{H}_{Static} \tag{3.19}$$

where v_{ab} is the absolute velocity.

In the convergent, the velocity of the fluid equals the difference of static and total enthalpies. In the divergent or diffuser we have a different formula due to the diffuser efficiency η, which comes from the easier detachment of flow:

$$\dot{H}_{Total} = \dot{H}_{Static} + \eta \frac{\mathrm{v}^2}{2}\dot{m}^2 \tag{3.20}$$

The absolute velocity has two components, axial and radial, where the axial component is linked to the mass flow by

$$v_{ax} = \frac{\dot{m}}{\rho A_{ax}} \tag{3.21}$$

Here we could introduce a blanking factor smaller than 1, to indicate that a fraction of the area is blanked or covered by the area of the blades. We omit this for simplicity.

Fig 3.21 gives the unwrapped active area of the gas turbine, that this along the circumference. Between the components and the rotor we have an interface infa, where the radial velocity changes according to

$$w_{tg} = v_{tg} + u \; ; \; w_{ax} = v_{ax} \tag{3.22}$$

Here, following gas turbine practice, v is the gas velocity relative to the stator and w that relative to the rotor. The quantity u is the velocity of the circumference of the rotor.

The principle is now to compute the radial force on each blade by the formula for aircraft wing lift. For N wings we obtain

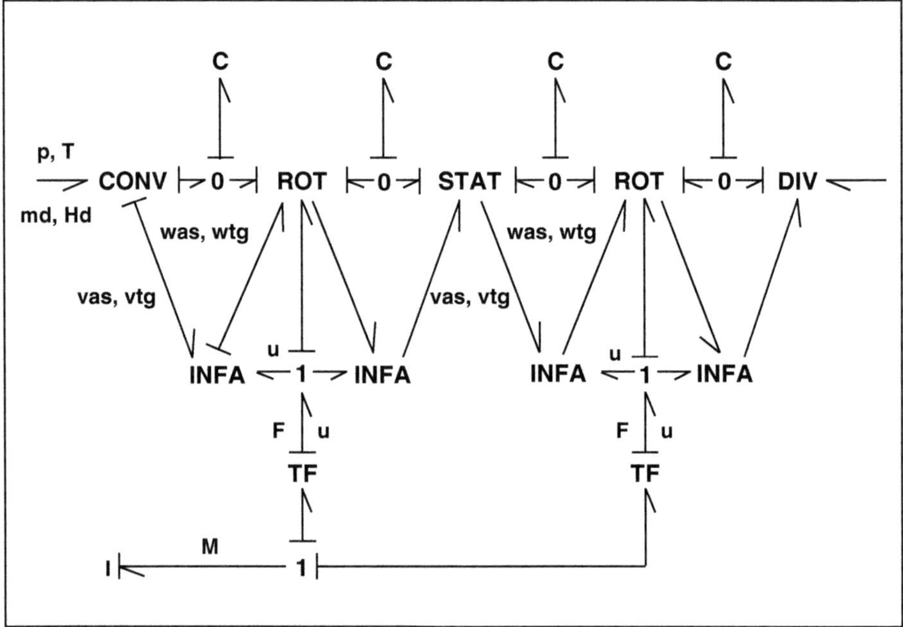

Fig. 3.22. Gas turbine BG with our symbols including an interface that transforms the velocity from a stator-referred part to a rotor-referred part by adding the circumferential speed u of the rotor.

$$F_L = N\frac{\rho}{2}\mathrm{w}_{ab}^2 A_w \mathrm{c}_{\text{lift}} \tag{3.23}$$

The radial outflow velocity can be calculated from a momentum balance, which comprises the lift force

$$F_L = \dot{m}\left(\mathrm{w}_{\text{tg1}} + \mathrm{w}_{\text{tg2}}\right)$$

which becomes

$$\mathrm{w}_{\text{tg2}} = \frac{\dot{m}}{F_L} - \mathrm{w}_{\text{Tg1}} \tag{3.24}$$

Fig 3.22 shows the BG, which is really a pseudo BG with two strands: one with pressure as effort and mass flow, the other with temperature and enthalpy flux. In addition we have the signal connections in the middle, where the passage between stator-referred and rotor-referred velocity takes place. The lower part of Fig 3.22, describes the rotary parts of the gas turbine, complete with transformers TF to go over to the rotary mechanical BG with I as moment of inertia. We have included the turbine example, even though it is not strictly thermodynamics, to show the power of the BG method when applied to such complex thermal fluid machines.

4

Chemical Reactions and Osmosis

4.1 Chemical equilibria and entropy stripping

We shall now extend our treatment to chemical reactions by means of a network of capacitors or multiport-C. As we shall see, they have a third bond with chemical tension as effort and molar flow as flow, whilst the other bonds, temperature and entropy flow and pressure and volume flow remain. So we deal with a true BG for chemical phases, or reactants and products.

First, we must be precise with the notion of equilibrium, because it is often treated too carelessly in the literature. For this we show in Fig 4.1 a boiler such as a reactor, with the steam or vapor phase above and the liquid phase below[1].

On the lower side a heat source can by applied to supply the heat for evaporation; we have already referred to this in section 3.5.

The phases are normally separated by gravity forces, but if they are unimportant, such as in clouds, both phases can be mixed[2].

In Fig 4.3 we show the system of Fig 4.2 as two multiport-C in equilibrium, where we actually have three separate equilibria.

1. Hydraulic equilibrium in the upper bond with pressure and volume flow: hydraulic bond;
2. Thermal equilibrium in the middle bond with temperature and entropy flow: thermal bond;
3. Chemical tension and molar flow in the centre: chemical bond.

---

[1] So the heat source is contained in the socket.

[2] There are even so called flash boilers (Benson boilers) for steam engines where the phases are not separated.

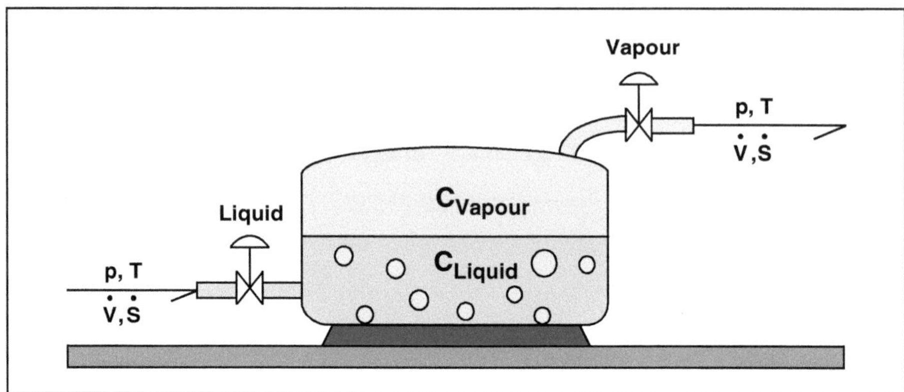

Fig. 4.1. Schematic of a boiler with steam and water phases. The top ensures that water and steam are at a constant pressure.

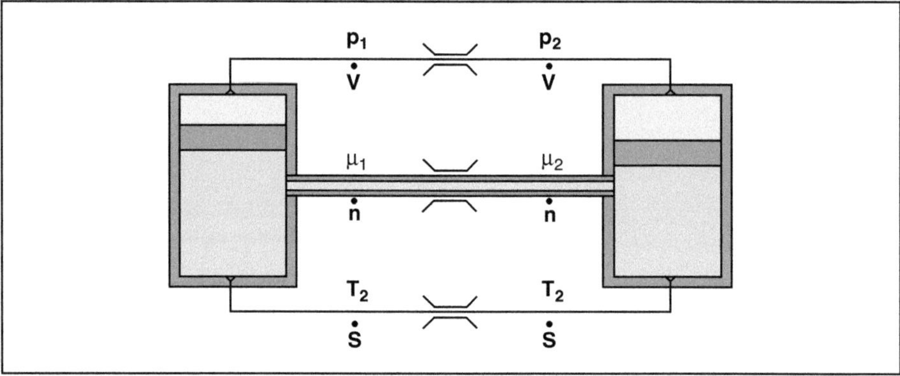

Fig. 4.2. Thermal, hydraulic and chemical equilibrium in two containers holding different substances. Between the two cells, there are three different exchanges.

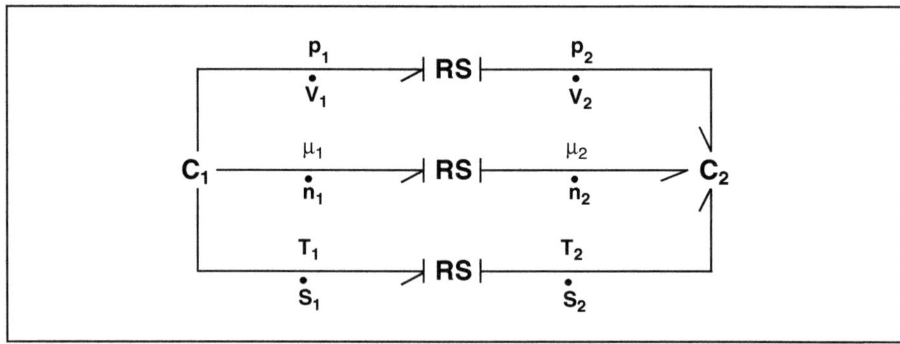

Fig. 4.3. Bond graph

In other words, we have not one but three equilibria between the phases in Fig 4.3 or between the phases in Fig 4.1. It means that volume flow can be exchanged by interface displacement, by entropy conduction over the surface and by a new set of variables, molar flow (as extensive variable) with chemical tension (as intensive variable).

$$\mu = \frac{\partial U(S,\ V,\ n)}{\partial n} \tag{4.1}$$

So all chemical equilibria are given by the equality of chemical tensions. A reaction runs downhill, one chemical tension (that of the reactants) decreasing and the other goes uphill, one chemical tension (that of the products) increasing until both are equal. In addition to the power entrained by the chemical bond, there will be a power exchange over the thermal bond, that is heat conduction, and over the hydraulic bond by a movement of the interface. These exchanges may have different time constants. As a special case, the equilibrium of the boiling surface and the changes it undergoes in a steam boiler can be taken as a chemical reaction with our three conditions of equilibrium.

As mentioned in section 1.3.4, normally one introduces a Legendre transformation here and writes

$$\mu = \frac{\partial G(T,\ p,\ n)}{\partial n}$$
$$G = U - T\,S + p\,V \tag{4.2}$$

Now the phases are considered to be homogeneous and the enthalpy density g is used, which is a function of T and p only[3]:

$$G = n\,g(T,\ p)$$
$$\mu = g \tag{4.3}$$

This form comes from the idea of homogeneity, where all is proportional to molar mass, usually expressed as the number of moles. We do not wish to say anything at this point about g(T, p), the density of free enthalpy.

For the second form of equation 4.3, we have to execute the derivation. Sometimes by exaggeration, chemical tension is written as being equal to the density of free enthalpy. This is a misnomer, because chemical tension is quite a simple variable with known dependence on pressure and temperature [JOB 1981] and [FUCHS 1996].

It is important to have a good understanding of chemical tension. According to equation 4.1, it is the derivative of internal energy with respect to moles, obtained by keeping entropy and volume constant. In other words, the mass

---

[3] Normally, one call this the free energy, but one should realize that it is not conserved because power can leave or enter through the thermal bond

transfer between phases does not take the entropy along; pure chemical tension alone is driving the reaction. So we shall illustrate this by the idea of entropy stripping.

Entropy Stripping

The key concept of entropy stripping (as well as the concept of volume stripping) is that entropy is stripped away before the reaction and reapplied to the products thereafter. Dissipated power is given by the difference of chemical tensions times molar flow

$$\dot{E}_{Dis} = \mu\dot{n} \tag{4.4}$$

Fig 4.4 represents a simple reaction with one reactant and one product, such as our liquid-vapor or water-steam equilibrium. It consists basically of two multiport-C, at left for the liquid and at right for the vapor. They are connected over a 1-junction (also called a series junction) to the central multiport-RS, which is the reaction resistance and which produces entropy. It is taken in conductance causality, hence it is rather a conductance: small conductance means a slow reaction.

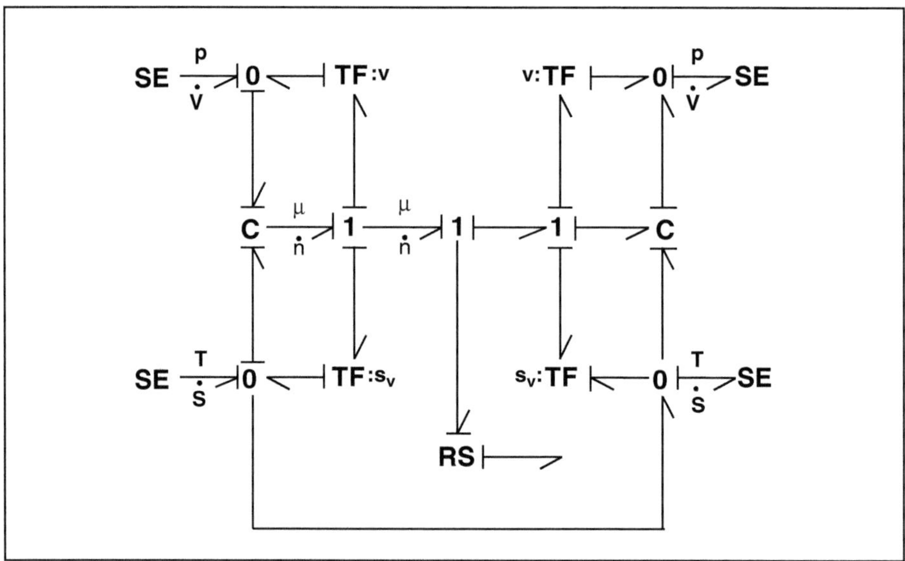

Fig. 4.4. Elementary chemical reaction (liquid-vapor equilibrium) as BG with entropy stripping. The transformer ratios are the entropy/mass density and the volume/mass density respectively. The bondgrapher will note that two effort sources on 0 junctions give a causal conflict. This is why we have not applied any causality to some bonds.

Entropy (and volume) stripping takes place in the various transformers, which have, as modulus, the entropy/mole (and volume/mole) ratio. The transformers naturally have two equations, one for efforts and one for flows. The effort equations give the passage through the TF from the density per mole and the chemical tension as follows:

$$\dot{S} = \dot{n}\ s; \quad \dot{V} = \dot{n}\ \mathrm{v} \tag{4.5}$$

The flow equations give an additional entropy and volume change expressed by:

$$\mu = u - T\ s + p\mathrm{v} \tag{4.6}$$

The relations 4.4 and 4.5 apply to reactant and product, or, as we shall see later (section 4.2), in chemical reactions to all reactants and products.

In the case of the liquid-vapor equilibrium, stripped entropy is the so-called latent heat of evaporation and the volume change, which are so important for steam. They are added to the conduction bonds through the parallel junctions. External effort sources for temperature and pressure keep their values constant under entropy and volume change. All this is reversible and the only irreversibility is in the central multiport-RS.

Reactants and products both experience the same temperature and pressure, therefore the effort sources should be combined into bonds linking to the environment. This applies both to stripped and unstripped entropy, and to volumes. They should be combined into reversible sources. The external effort sources for temperature and pressure both impose effort causality, as is the property of all sources. Therefore, the main multiport-C for reactants and products go into derivative causality.[4]

The remedy is simple: interpose small R-elements between the effort sources and the first junctions, as shown in Fig 4.5. They are purely dummy and have no effect on the system dynamics. Indeed, this remedy is frequently used in electrical engineering [GDT 2000] and also in mechanical engineering to avoid derivative causality. In our experience, this is more efficient than trying to handle derivative causality using computer programs.

Another dummy RS is needed in the connecting bonds between the multiport-C for liquid and vapor, in order to couple them thermally together, an aspect which is unavoidable. These dummies should have very large conductance, such that the time constants with the C-elements and multiport-C are about a million times smaller than the other time constants of the BG. Thus, in

---

[4] There is no escape from the fact that connecting an effort source with a C-element leads to derivative causality as does the connection of two C-elements in parallel.

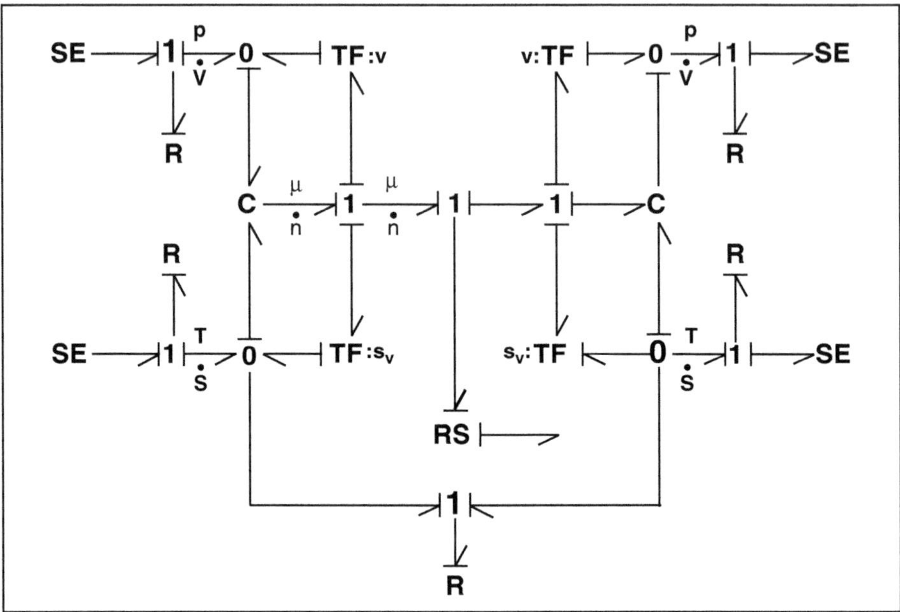

Fig. 4.5. External SEs and dummy resistor added to the preceding figure to avoid derivative causality. The sources, temperature and pressure are common to all reactants and products.

effect, they compensate for small differences of the multiport-C by large compensating currents or flows.

Note that in Fig 4.5 all elements including stripping and unstripping are reversible, but not the dissipation in the central RS multiport. We have two entropy flows:

1) the flow coming out from the central RS dissipation and

2) the net entropy flow coming from the transformer TF at the lower left side and the unstripping from the transformer at the lower right side.

This also explains why endothermic or heat absorbing reactions exist: if the products have greater entropy capacity than the reactants, the net stripped flow is negative. If this exceeds the dissipation, the deficit must be supplied by the environment. So the reaction absorbs entropy that tends to cool the environment.

Some humorists call the whole procedure the striptease of entropy and liken it to a girl dancing in a cabaret: she takes off her clothes before the act and puts them on again afterwards.

Once, an author of a thermodynamic textbook said that in order to keep entropy and volume constant, i.e. to realize equation 4.1, one has to set temperature and volume high for reactants and low for products. To repeat, we call this the taking away of entropy and volume, entropy stripping or the striptease of entropy.

Unfortunately by going to free enthalpy, the power relations of equation 4.2 are obscured because the temperature and pressure sources themselves absorb power, as we have seen in chapter 1. So free enthalpy power is no longer conserved. Hence a proper treatment of power and energy is no longer possible with the form of equation 4.2, and we will not use it.

Our approach gives many new insights into chemical reactions. In particular, as we have seen, it can explain why some chemical reactions are endothermic, that is they produce cold and not heat, whilst most produce heat. And for this we need the concept of entropy stripping explained above in this section.

We continue, in our treatment of chemical BGs, to simplify our Fig 4.5 and Fig 4.6, because we can still see reversible stripping and unstripping using the coupler element "coupl", which replaces one 1-junction and two transformers. We can also see the irreversible dissipation due to the main chemical friction. In going to the environment, both entropy flows are mixed and make up the enthalpy of the chemical reaction, which really comes from the chemical

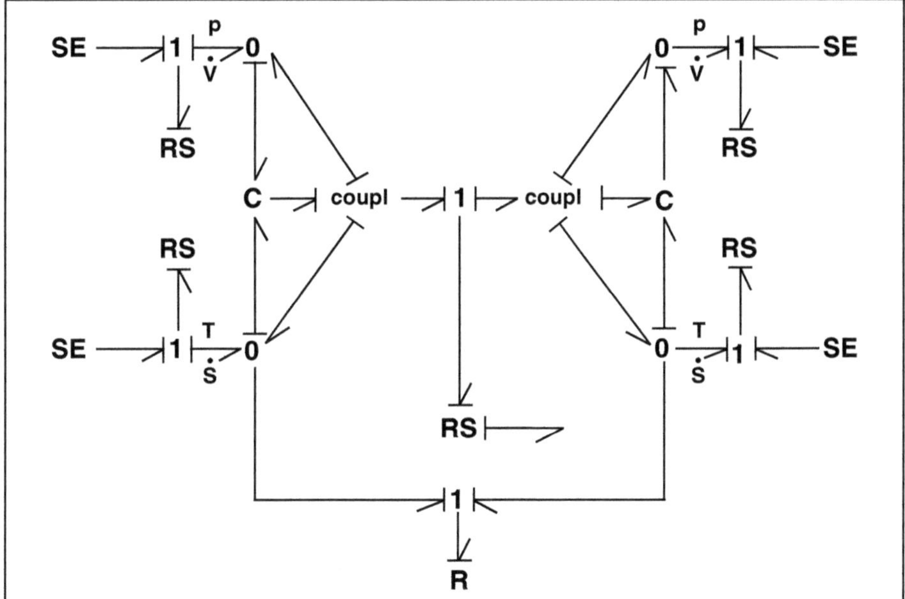

Fig. 4.6. Simplified reaction, where the transformers and their connecting 1-junctions of the preceeding figure are replaced by the coupler coupl elements.

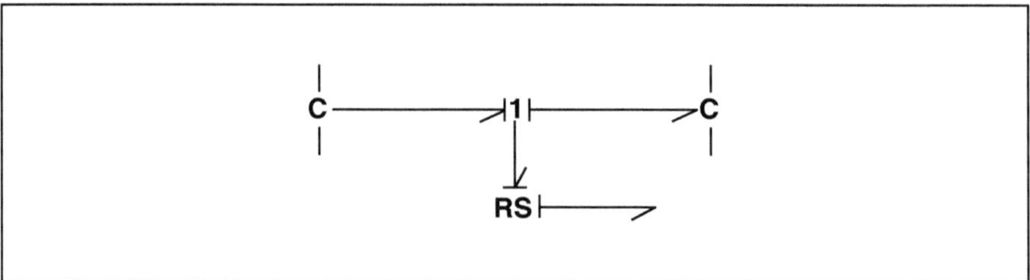

Fig. 4.7. Kernel of a chemical reaction with one reactant and one product, and the RS producing entropy.

bonds combined. Here again we have introduced the dummy resistor R to obtain proper, integral causality.

The reversible entropy of Fig 4.7, as obtained from stripping and unstripping, is summarized by vertical short strokes. This simplification is used for the chemical reactions of the next section. If we further eliminate the S part of the chemical resistances, we obtain the chemical BG as used in the literature [OPK 1973]. In this sense the chemical BGs in the literature are incomplete, but may be sufficient if one is not interested in thermal effects assuming constant temperature operation. But these effects can exist and generally make themselves felt.

The entire entropy and volume stripping reduced to the small vertical bonds on the multiport-C will be used for the more complex reactions in the next section. In the literature, OPK eliminate the couplers and conduction bonds altogether. Naturally they exist and may produce unexpected effects.

Contrary to fluid in tubes and components, we have here no spatial dependency but a dependency between two or more chemical phases. The chemical literature eliminates the spatial dependency by postulating a CSTR (Continuous Stirred Tank Reactor). As mentioned, we have encountered this phenomenon already with boiling and two-phase flow in Section 3.5.

Hence we have also eliminated all spatial dependency and can consider in the next section only a reaction between several reactants and products.

To illustrate the entropy capacity of matter as chemical reactant, we give the example of wine glasses, as originally adduced by Wiberg (1972). Entropy capacity is shown on the horizontal axis and temperature on the vertical axis, the curves of entropy capacity having the form of a wine glass, as shown in Fig 4.8. The contents are the total entropy if the glass is filled to a certain temperature level. The figure shows two glasses of different diameters or sizes.

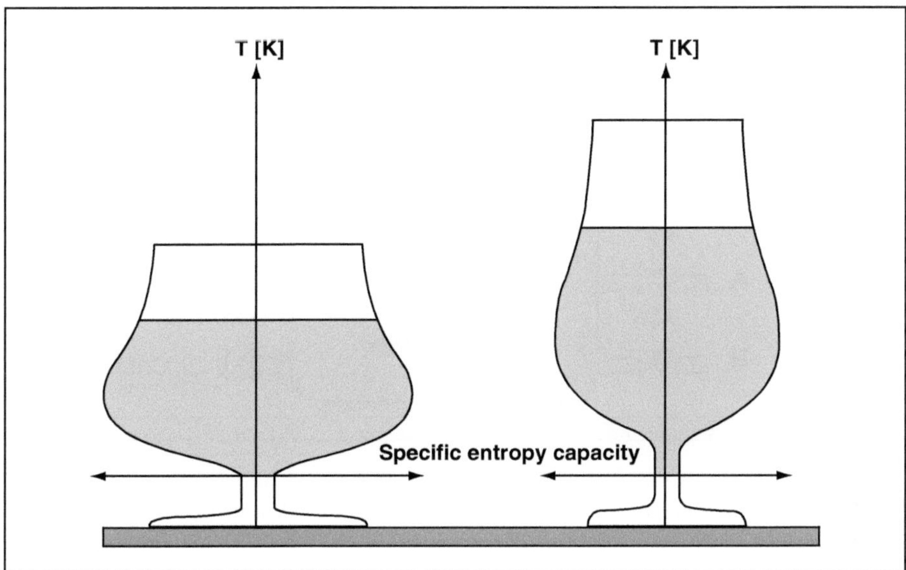

Fig. 4.8. Temperature/entropy curves in the form of two wine glasses. The capacity of structural entropy is like the capacity of a wine glass, with the diameter and the form changing by the reaction. Varying the form changes the volumes. Therefore entropy must be supplied from the environment to keep temperature constant or, if the glasses are insulated, temperature must vary. This is the explanation why, with entropy capacity increasing, temperature is decreasing, giving endothermic reactions (after Wiberg).

The phase change or chemical reaction can be visualized as a deformation of wine glasses. In Fig 4.5, going from left to right, entropy capacity is decreased. Since total entropy cannot be destroyed, the temperature level increases when the volume becomes smaller, that is we have a normal exothermic reaction, to which the entropy of dissipation still comes. On the other hand, if the entropy capacity is greater in the products, the temperature level will decrease and entropy must be supplied from the environment. In this case we have an endothermic reaction. As mentioned, a liquid-vapor equilibrium can be thought of as a reaction with one reactant and one product. The difference in entropies becomes the latent heat of evaporation, which depends on the temperature.

The picture with the transformation of wine glasses was conceived by E. Wiberg, and we regard it as a good illustration of structural entropy and temperature that includes the influence of a change, i.e. a chemical reaction. His book is also remarkable in that he considers entropy as thermal charge, as we do.

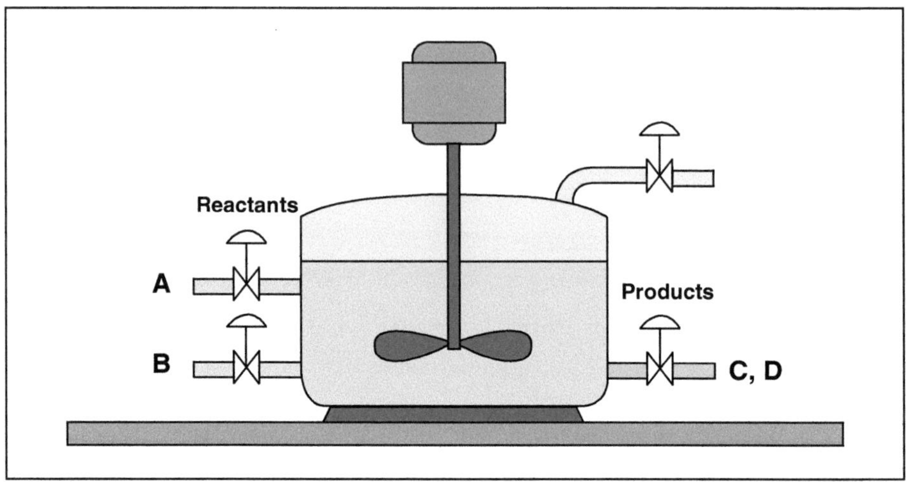

Fig. 4.9. Reactor vessel, well stirred, with 2 reactants A and B, and 2 products C and D.

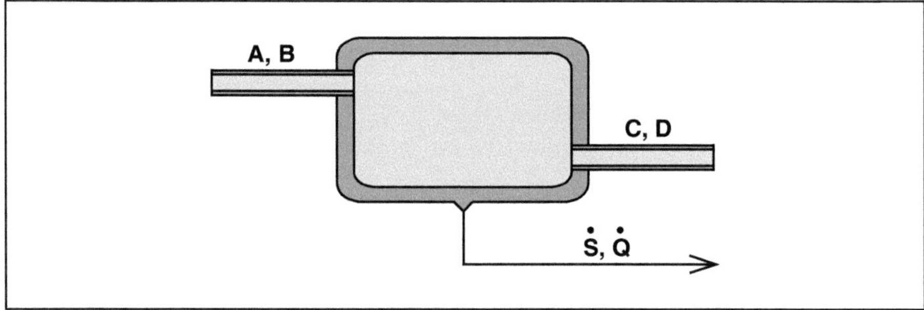

Fig. 4.10. Schematic.

4.2 Chemical reactions

We now extend our BG to more complex chemical reactions, say with two reactants and two products as shown in schematic in Fig 4.10.

To fix ideas, Fig 4.9 shows a reactor vessel, well stirred, called in the trade Continuous Strirred Tank Reactor CSTR, for two reactants and two products[5].

There are two inlet pipes for the reactants and one outlet pipe for the products. The chemical or stochiometric equation is in this case

$$\nu_A A + \nu_B B = \nu_C C + \nu_D D \tag{4.7}$$

---

[5] A heater in the socket is optional.

which represents a chemical reaction with two reactants and two products, each associated with a multiport-C. The coefficients

$$\nu_A, \quad \nu_B, \quad \nu_C, \quad \nu_D$$

represent the so-called stoichiometric coefficients, simple integers.

Fig 4.11 and all the following figures have the simplification of Fig 4.7. The whole stripping and unstripping process is shown below. We remind the reader that we have two entropy (and heat) flows: reversible stripping and unstripping entropy is indicated summarily at the bottom.

The irreversible dissipation is in the twoport RS.

To repeat, we have shown entropy and volume stripping in figure 4.11 by small vertical bonds on the multiport-Cs and added it below only by the lower parallel 0-junction for entropies. It is combined at the upper parallel junction to connect with the environment, represented by an effort source SE.

There are no difficulties with causality in this form (add resistors for external SE, as with boilers). Also the series junctions could be combined, but we have left them separate in the lower part of Fig 4.11, as mentioned, in order to show the structure of the reaction.

The transformers are important. They show the relation between chemical tensions and the so called affinities A, which correspond to the stoichiometric coefficients ν in equation 4.7. The effort equation introduces a new variable,

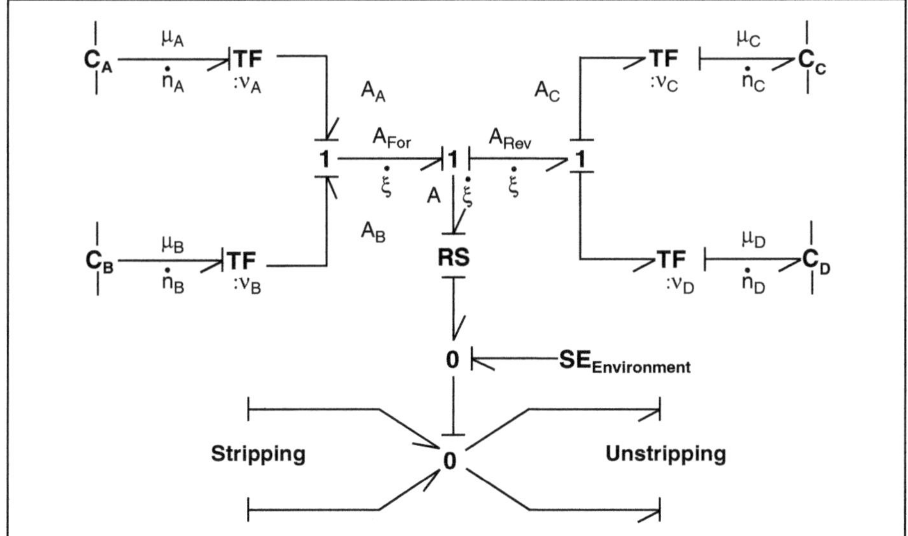

Fig. 4.11. BG for a simple reaction between two reactants and two products, where the transformers are equal to the stoichiometric coefficients ν.

chemical affinity, calculated from the chemical tensions. The flow equation for these transformers gives the relation between the progress of reaction $\dot{\xi}$ and the flows of reactants in the multiple-C.

Expressed as an equation we have for the transformers

$$A_i = \nu_i c_i \quad where \quad i = A, B, C, D$$
$$\dot{\eta}_i = \nu_i \dot{\zeta}_i \tag{4.8}$$

Hence chemical affinity A is an effort variable which represents the force driving the reaction. It's complement $\dot{\xi}$ is called the advance of the reaction. For this rate, the following equation has been given

$$\dot{\xi} = \dot{\xi}_1 + \dot{\xi}_2 - \dot{\xi}_3 - \dot{\xi}_4 \tag{4.9}$$

which is correctly represented by the central series 1-junction in Fig 4.11.

For the chemical tension in the C-multiports one takes (Denbigh 1971, Thoma et al 1977)

$$\mu_A = \mu_{A0} + R\,T \log(c_A) \tag{4.10}$$

with μ_A = chemical tension and μ_{A0} the concentration independent part thereof, R = gas constant, T = absolute temperature, and the same equations for B, C, and D.

Inserting equation 4.10 into 4.9 gives

$$\dot{\xi} = \dot{\xi}_0 X \left[c_A^{\nu_A} c_B^{\nu_B} - c_C^{\nu_C} c_D^{\nu_D} \right] \tag{4.11}$$

where the constant X lumps together all other parts, especially the concentration independent parts of chemical tension; all the chemical tensions depend on temperature. Chemical equilibrium is obtained if the bracket vanishes in equation 4.11. We can define a distance from equilibrium by

$$K(p,\,T) = \frac{c^A c^B}{c^C c^D} \tag{4.12}$$

where, as indicated, the distance depends on pressure and temperature (see [THOMA 1977]) for details.

Equation 4.12 is the well known law of mass action. It is remarkable that this law follows directly from our Bondgraph representation.

4.3 Near to and far from equilibrium

According to the Bondgraph in Fig 4.11, it is the difference of forward affinity A_f and reverse affinity A_r that drives the reaction, the speed of which is controlled by a multiport-RS behind a series 1-junction. If forward and reverse affinity is equal, the resulting affinity A vanishes and one speaks of chemical equilibrium, as previously stated. It should also be noted that we have one degree of freedom (DOF) for the reaction.

In certain cases, the reaction rate is more complex and depends not on the difference but on the forward A_f and reverse A_r affinities individually; this is the definition of "far from chemical equilibrium".

This definition takes an important role in biology and, in Bondgraphs, it is simple to incorporate with the separate variables as in Fig 4.12 below. Let us repeat: in the literature this is called far from equilibrium. In all reactions, multiple-Cs are functions of temperature and pressure. This is also the case for multiport-RSs which behave almost like a switch: above a certain temperature, their reaction starts.

Note that we still have the same mass flow but two affinities, in other words only one DOF.

Basically, reaction kinetics is contained in these multiport-RS, and the so-called parallel and competing reactions are explained in terms of these elements (see next section).

To develop the chemical kinetics further, one would use the expression

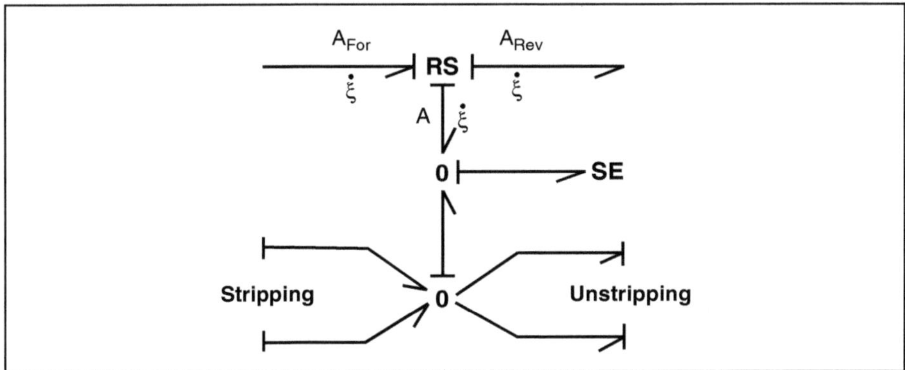

Fig. 4.12. In certain cases, far from chemical equilibrium, the reaction rate is not equal to the difference of affinities, but depends on them individually. Below, we show the stripping and unstripping. The flow experienced at the SE is the observable entropy of reaction.

$$\dot{\xi} = F\left(\frac{A_f}{k\,T} - \frac{A_r}{k\,T}\right) \tag{4.13}$$

near to the equilibrium, and

$$\dot{\xi} = F\left(\frac{A_f}{k\,T},\ \frac{A_r}{k\,T}\right) \tag{4.14}$$

far from the equilibrium

In [THOMA 1977-2], authors endeavored to obtain a good Bondgraph for reactions far from equilibrium. We feel now that the best representation is our Fig 4.12 simply with two separate bonds for A_f and A_r affinities, and we note separately that the mass flows are the same.

4.4 Parallel and competing reactions

In chemical reactions, there can be competing reactions that produce new products from one of the reactants.

Fig 4.13 shows the mechanism of parallel and competing reactions.

For the competing reaction there is the 0-junction from which some of product B is withdrawn. It combines with a new reactant E to form a new product F in a competing reaction. An important feature is that the reaction has another multiport-RS, which can have a different temperature dependency so that it is operative only at higher temperature. This can explain effects such as food burning. The burning reaction is controlled by another RS which becomes operative only at higher chemical tensions. In a sense, the burning reaction switches on only at high temperatures[6]. Parallel reactions are similar. They also have different reactants with another multiple-RS to make the same product, or a mix of products. We repeat that the stripping and unstripping of entropy and volume exist in all chemical reactions along with their influence on temperature. As a consequence, with more complex reactions many possibilities of oscillation exist as in the biological system in Fig 4.14.

4.5 Osmosis as a bridge between physical, chemical and biological effects

Osmosis is halfway between physical and chemical phenomena and applies to biological phenomena, a particular case of life sciences. In this connection, BG

---

[6] A good illustration is the burning of food in the kitchen: too high a temperature opens, by the temperature dependance of an RS, a competing reaction turning the food black.

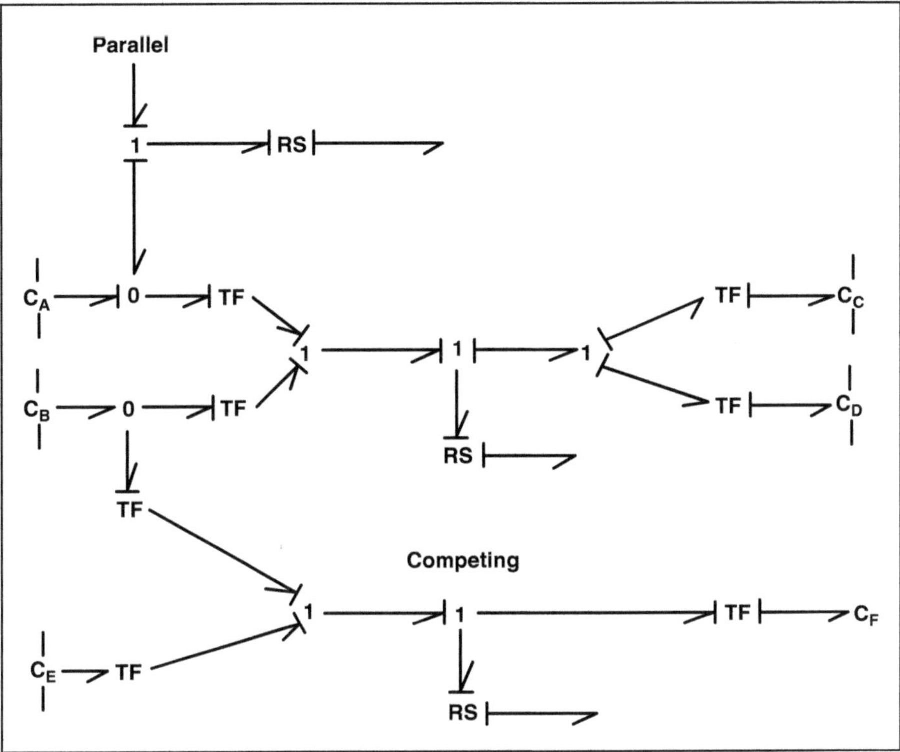

Fig. 4.13. BG with parallel reaction on top and competing reaction from E to F below.

can be applied there and especially by the use of the notion of tensor BGs [MOCELLIN 2001].

Osmosis describes a pressure generated between two compartments separated by a selective membrane (sometimes called a semipermeable membrane). The repartition of particles result from the selective permeability of the membrane: the distribution of water depends on osmotic pressures across the membranes.

Fig 4.15 shows in a much bigger arrangement ($\approx 10\,\text{cm}$) the principle of two osmotic compartments separated by a selective membrane. The inner cell is filled with pure water and the outer vessel with a solution of salt and water. The membrane lets the pure water pass through, but not the salt.

We have, as mentioned, the three equilibria: hydraulic, thermal and chemical. Consequently the pressure on the solution side is higher.

In osmosis there is a contradiction between hydraulic and chemical equilibrium: for membranes with very small pores (about $100\,\text{nm}$), the chemical equilibrium for pure water applies. The water passes though the membrane

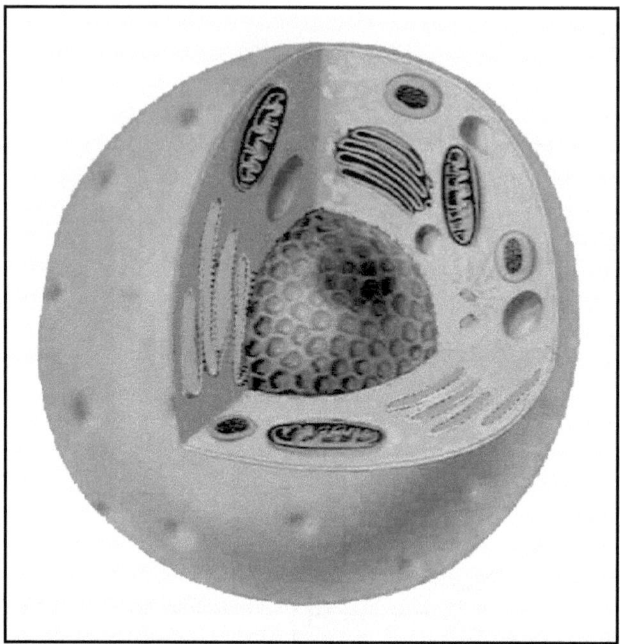

Fig. 4.14. Biological cell where we have selective membranes everywhere: outside the big sphere and also inside between the cell and the several compartments. The cell is very small, about 800 nm in diameter.

following chemical equilibrium and therefore raises the hydrostatic pressure in the water and salt solution.

With somewhat larger pores we have a double effect, where both chemical and hydraulic equilibrium applies; this is known as a leaky membrane (Fig 4.16). It is derived from the entropy stripping with two multiport-Cs separated by a twoport-RS (top center).

Fig 4.16 shows osmosis by BG, which is used for sea water desalinisation and everywhere in nature for biological processes. The main features are two multiport-C for sweet and salt water, or for the exterior and interior of a living cell. They have the same structure as those of chemical reactions with three efforts: temperature, pressure and chemical tension, and three flows: entropy flow, volume flow and molar flow. The main feature is the resistance in the center, with the difference of chemical tension that produces the molar flow and necessarily new entropy. Hence it is irreversible. So we have a very simple chemical reaction with one reactant and one product. In other words the transformers with the stoichiometric coefficients of chemical reactions become equal to one and disappear.

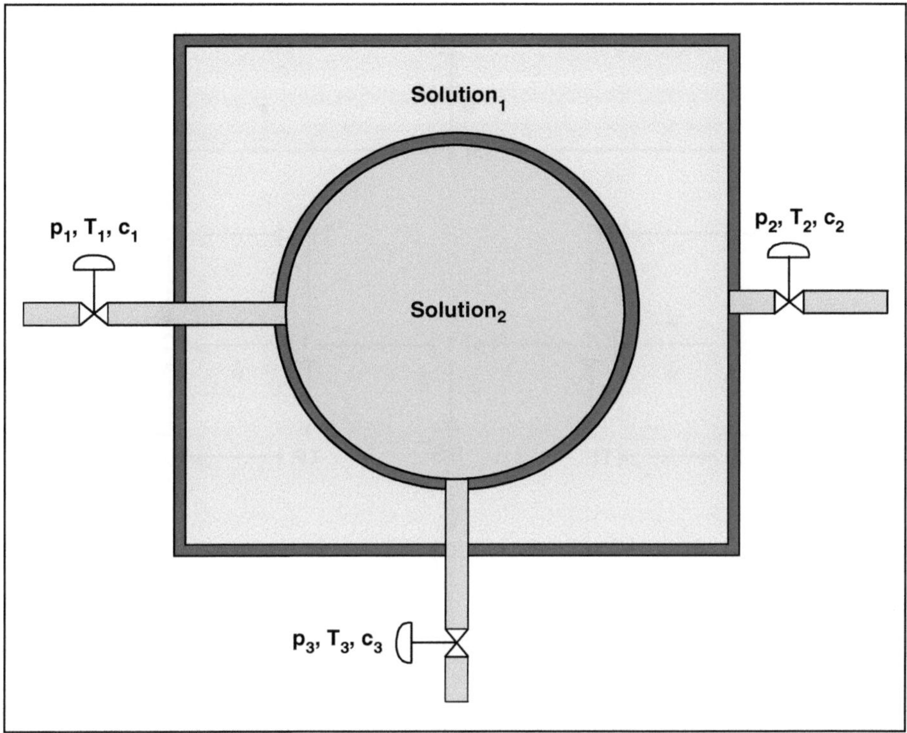

Fig. 4.15. Vessel with two compartments separated by a selective membrane. One has water and salt, the other pure water. The water pressure goes higher at the pure water side because in such membranes chemical equilibrium applies, leading to a higher pressure in pure water.

We have a second twoport-RS (top center) which is sensitive to the pressure difference.

The net effect of entropy stripping is the lower bond which adds entropy to the new entropy from dissipation by a parallel junction and which appears and is measurable at the environment. This replaces the temperature-entropy sources from before.

The bonds with pressure and volume flow on top connect over an RS-multiport. This is normal volume flow in a leaky membrane and the dissipation produces new entropy, but can often be neglected.

The lower part of Fig. 4.16 shows the sources extra with combined dummy R-element

Osmosis is treated by Thoma-Bouamama in [GDT 2000], Fig. 6.24 and 6.25 and by Thoma in [THOMA 1985].

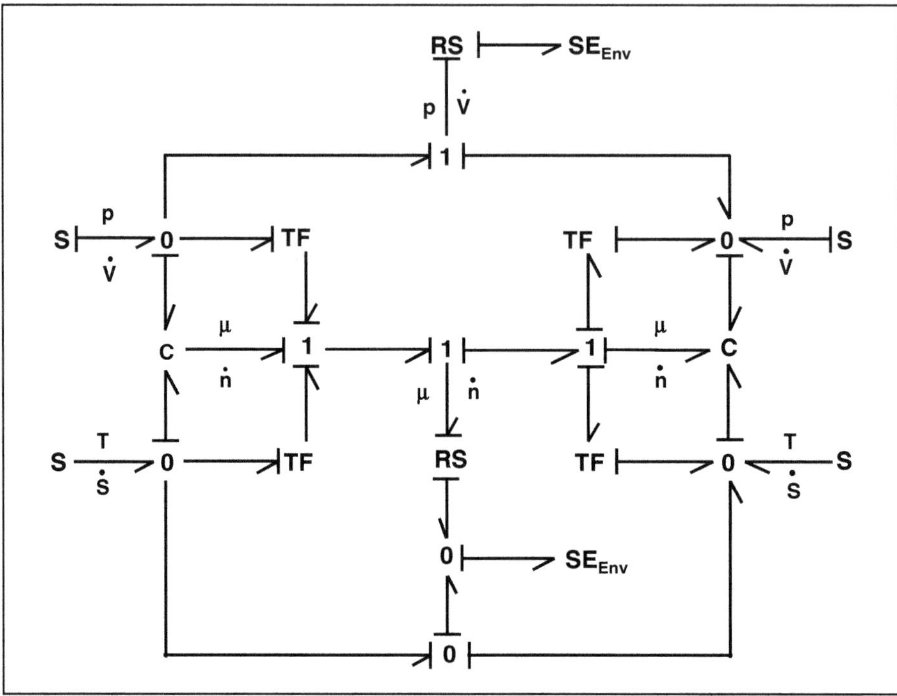

Fig. 4.16. BG for osmotic pressure generation between vessels of Fig 4.15. We have two twoport-RS here, one (above) sensitive to chemical tension and the other (below) to hydrostatic pressure: this leads to a leaky membrane.

4.6 Reversible mixing and Gibb's paradox

Mixing and diffusion are on the list of entropy sources, but this only applies to simple or unobstructed mixing. There exists also a reversible mixing that generates mechanical energy but no entropy; we shall describe this here. To construct the process we must use the concept of selective walls or pistons, which we have used already in section 4.5. So if a piston allows gas A through and obstructs gas B, then the force is cross-section times partial pressure of gas B, while gas A goes though the piston unimpeded and exerts no force.

In Fig. 4.17 both gas are separated and the pistons are close together. Next we imagine the cylinder with the two pistons separated, the left one allowing gas A to pass and stopping gas B, and the right one allowing B to pass but not A. This is shown in Fig 4.18. In the space between them both gases are present, i.e. the gases are mixed.

Hence we have made an entropy conserving mixing process, but the gases become colder with expansion into their larger spaces. The pistons undergo forces, proportional to the blocked partial pressures, which must be taken

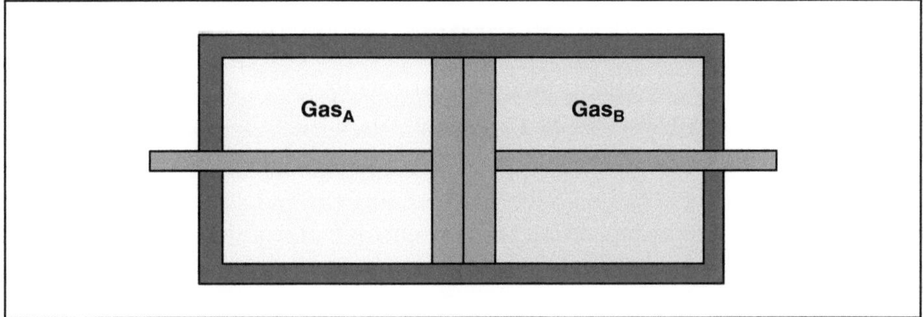

Fig. 4.17. The selective piston in left vessel lets gas A pass freely and blocks gas B. Hence the force is given by the pressure difference of gas B times piston area, whilst gas A has no influence.

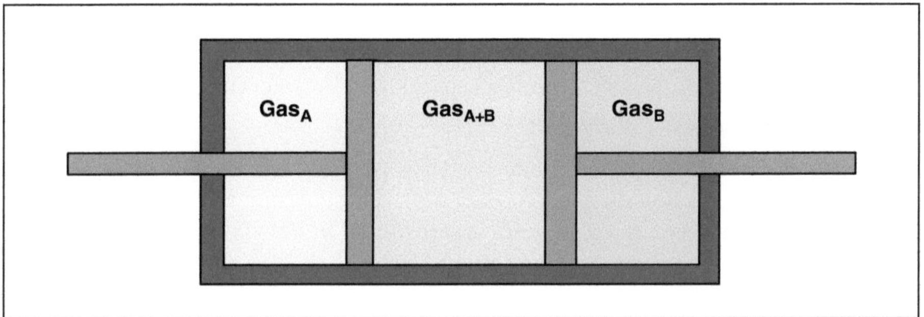

Fig. 4.18. Reversible mixing: Two pistons in a vessel, one permeable for gas A the other for gas B. They experience different forces that are taken up by a brake, or rather by a reversible hydraulic motor. This generates entropy on mixing. If one returns it then one obtains the same entropy by bursting a separate membrane.

up by suitable brakes, and this produces entropy as shown by multiport-RS. If this entropy were to be used to reheat the gases (not shown), we would recover pressure and entropy on them. In so doing we would return to the case of simple, entropy producing mixing.

Remember that each piston feels the partial pressure of the gas it blocks and transmits it over a pushrod to an external force generator (not shown) and that he pushrods regulate the movement of both pistons.

To repeat, piston A receives the partial pressure of gas B, and piston B the partial pressure of gas A. So going from Fig 4.17 to Fig 4.18 produces a gain in mechanical energy, which can be dissipated by a suitable brake. If a hydraulic (or other type of) motor were to be used instead, the device would be completely reversible and no entropy would be generated: reversible mixing.

If a brake were to be used, the process would be irreversible and entropy generating. Also the gas cools on expansion. If the entropy generated were to be re-conducted to the gas, it would return to its exact former temperature. Then we would have the case of irreversible or simple mixing.[7]

Summarizing, Fig 4.18 gives an apparatus for reversible mixing and unmixing, simple mixing being obtained merely by running the gain in mechanical energy through a brake and reconnecting the new entropy. Returning to simple mixing by bursting membranes would generate entropy that could in principle be recovered by selective pistons.

A frequently asked question is what happens if the gases become very similar. Some authors say that it is an irregularity of nature if there is the slightest difference between two gases: there is mixing entropy or else there is not. Quantum mechanics assures us that there is always a finite difference between gases or none at all. This is called the Gibbs paradox (Prigogine and al. 1998, page 155).

We have a different attitude which does not involve quantum mechanics: different gases can always be separated reversibly by selective pistons, and this produces entropy only by external dissipation.

So the problem is to find a selective piston, which becomes more difficult for increasingly similar gases. We also feel that entropy is always calculated from a certain reference state and its absolute value has no significance. If better selective pistons were invented, this would change the absolute value of entropy: an impossible state of affairs. So it is only the reference values that count.

One could object that we have the Nernst theorem (3rd law of thermodynamics), which states that all entropy content vanishes at zero temperature. This is only a theoretical law, because entropy at very low temperature is not accurately known.

Also one would never know whether or not entropy is frozen into certain substances. Also interesting is superconductivity, especially the fact that Helium splits into two phases, He I and He II at a temperature of about 4 K. Phase

---

[7] The Fig. 4.17 and 4.18 have a certain resemblance to Fig 5.4 of R. Feynman (from T. Hey and al., the Feynman Lectures on Computation, Perseus Publishing, Cambridge, USA, 1999). He uses two pistons to displace an atom from one equilibrium state to another. There, he speaks about computation with single atoms or even with a single degree of freedom of atoms. Therefore he descends size, as we shall mention at the begining of chapter 5, by about 20 orders of magnitude and thinks that the applicable laws are still the same. So for us these pages contain interesting thought experiments that need to be proven. Nevertheless, the chapter on Thermodynamics of Computation, and indeed, the whole book are worth while reading.

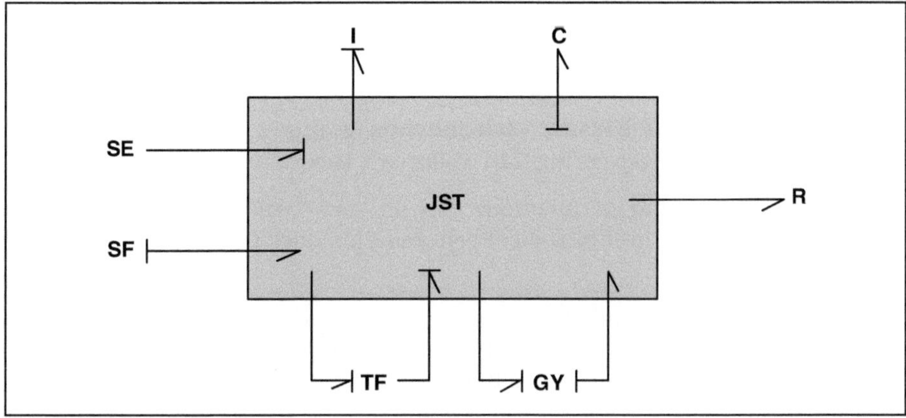

Fig. 4.19. A Bondgraph with all junctions inside a conceptual boundary and all other elements outside it. Bonds penetrating the boundary are labeled 1 to i and Tellegen's theorem applies to them.

II seems to have no entropy and this leads to many strange effects near to absolute zero temperature[8].

To summarize, we consider entropy as zero at the reference temperature and pressure, and take all values from there.

4.7 Tellegen's theorem and chemical inductance

There are two further points pertaining to Bondgraphing that appear important to us.

Tellegen's theorem:

This theorem was stated back in 1946 for microwaves and adapted to Bondgraphs and chemical thermodynamics by Atlan-Kachalsky in 1973.

For Tellegen's theorem, we place all junctions inside a conceptual boundary and all other elements outside it, as shown on Fig 4.19, so we have a junction structure JST. The bonds penetrating the boundary are labeled 1 to i. Tellegen's theorem applies to them and states that the sum of powers is conserved:

---

[8] Falk (personal communication to Thoma) told that the Nernst theorem is like counting altitudes in geography. In principle altitudes should be counted from the center of the earth since that would be more correct; but in practice sea level is taken because it is known with greater accuracy.

$$\sum_{1}^{i} e_i\left(t\right) f_i\left(t\right) = 0 \tag{4.15}$$

This is obvious to us because each junction is power conserving, hence all junctions are power conserving and Tellegen's theorem follows.

Note that the number of junctions can be very large, of the order of the number of molecules in a biological cell, but this does not impair the validity of Tellegen's theorem.

Less obvious is the extended version in which power is also conserved. Equation 4.16 means that the product of efforts at one point in time with the flows at a later time is also conserved. Tellegen called it quasi power.

$$\sum_{1}^{i} e_i\left(t\right) f_i\left(t - \tau\right) = 0 \tag{4.16}$$

τ = time delay

Further, Atlan told Thoma that it also applies to variable systems, as long as the junction structure (inside the boundary) remains constant. This is important in biology where the outside elements can change as an organism grows. He thought specifically about biological applications where the number of penetrating bonds is very large, something equating to the number of molecules in a living cell, or 10E23.

Chemical Inductance

Normally, thermal and chemical networks and Bondgraphs have no inductance.

However, one publication exists [ATLAN 1973-1] in which time delays in chemical reactions are expressed by inductances. This is obtained by a Taylor expansion of the chemical flow and breaking off after the second terms, which can be likened to an inductance.

This example is interesting because it shows inductance in chemistry, which some people have said cannot exist because it contradicts the second law of thermodynamics. This was shown to be an error [THOMA 2000], because the second law applies only to steady state, and says nothing about transients. It is not very important in itself, but illustrates fine points of the second law.

We would conclude this chapter by saying that, even in chemistry and by extension in biology, understanding is much improved by the use of Bondgraphs. The most important points are endothermic and exothermic reactions, which Jean Thoma never understood until he wrote a Bondgraph.

Another, more recent application is osmosis, of which H. Atlan did not previously have a full understanding inspite being a professor of biophysics. He helped to draw up a Bondgraph, which is shown here in section 4.4.

5

Entropy and Information Theory

5.1 Orders of magnitude and microscopic entropy

So far we have offered a macroscopic theory of entropy and considered it as a gray paste. Now we wish to examine the statistical aspects and establish the connection with the well-known formula of Boltzmann and with information theory in communication and electronics.

Let us pause for a moment and consider the change of viewpoint. Our thermodynamics was originally concerned with machines of a certain size, as were the BGs to describe them. Now we wish to go to the size of single atoms or even to the DOF (degrees of freedom) of single atoms. So we descend about 20 orders of magnitude (10E-20) and hope that the laws of nature there are still the same.

All matter is made up of atoms or molecules and each has several DOF. Each can be in one of several states with a certain probability, and each state has a probability of i. We only know that the DOF is somewhere, and consequently the probabilities of all states add up to one:

$$\sum_i p_i = 1 \tag{5.1}$$

With this comes the famous formula of Boltzmann:

$$S = -ka \sum_i p_i \log p_i \tag{5.2}$$

where a is a normalization constant for changing from log base 2 to log base 2.71.

Note that the probabilities are constrained by the total energy, which in turn is given by the so-called equipartition theorem of Boltzmann:

$$E = kT = \sum_i \varepsilon_i \, p_i \tag{5.3}$$

$$\varepsilon_i = Energy \ of \ DOF \ i$$

This is then summed for each DOF and atom to obtain the entropy of matter. Note also that all probabilities are smaller than one, thus the logarithm is negative and the whole entropy positive with the minus sign in front.

Important also is the fact that the contribution to each state becomes very small or zero if:

1. The probability p_i is very small and we have a large number of states, each with vanishing probability, under the above energy constraint;

2. The probability p_i equals one, and all others are zero. This means we have certainty that the DOF is in this state and the logarithm vanishes. This is necessary at zero temperature to keep specific heat finite, or, with quantum theory, equal to zero.

Contributions to each DOF are only in the middle range, where some of the states have a certain significant probability.

Inspired by quantum mechanics, one can say that each DOF contributes a quantum of entropy. However, this is not constrained to integral values, but pondered by the Boltzman formula[1]. The next point is that for macroscopic entropy we have to sum over a very large number of atoms (Avogadro's number), with each multiplied by the number of DOF for each atom.

One question that has been asked is how Boltzmann entropy and macroscopic entropy are related. To this is added the entropy of a message after Shannon. In the authors opinion, they are in principle the same, but distinguished numerically. Only the magnitudes differ significantly, as mentioned, by the order of the number of DOF, that is Avogadros number.

It is still unclear how the principle of conservation of macroscopic entropy can be derived from the Boltzman formula. The main point is the increase in entropy by conduction, where entropy flow becomes greater while heat flow remains constant in passing from higher to lower temperature.

5.2 Entropy of a message after Shannon

Information after Shannon is the irreducible content of a message, either by telegraph, telephone or radio. Each message consists of many signals that are transmitted consecutively and each signal can take several values with a probability p_i. The index i runs from 1 to n and we have the probability normalization

$$\sum_{i=1}^{n} p_i = 1 \tag{5.4}$$

---

[1] Thoma discussed this in 1985 with Falk.

The information carried by a certain signal is given by

$$\mathrm{ld}\ p_i = \frac{\ln p_i}{0.693} \tag{5.5}$$

The logarithm to the base 2 or ld is chosen because a signal with a probability of 0.50 gives one unit of information, commonly called one bit. The mean information carried by this symbol is

$$p_i \mathrm{ld}\ p_i$$

and the mean information per position in the message is

$$S = -\sum_{i=1}^{n} p_i \mathrm{ld}\ p_i \tag{5.6}$$

In a simple example, let the signal have only two values (n = 2) with $p_1 = p$ and $p_2 = 1 - p$, which gives us

$$S = -p\ \mathrm{ld}\ p - (1 - p)\mathrm{ld}\ (1 - p)$$

In the special case that p = 0.5 we obtain an entropy of one, that is an entropy transmission of one unit.

Fig 5.1 gives an example of a communication system. Information is stored at left in a memory, read by a scanner and sent by a transmission channel to a receiver with its own memory.

In the memory of the receiver there are several elements, each having a probability i and an entropy

$$S = p_i \mathrm{ld}\ p_i$$

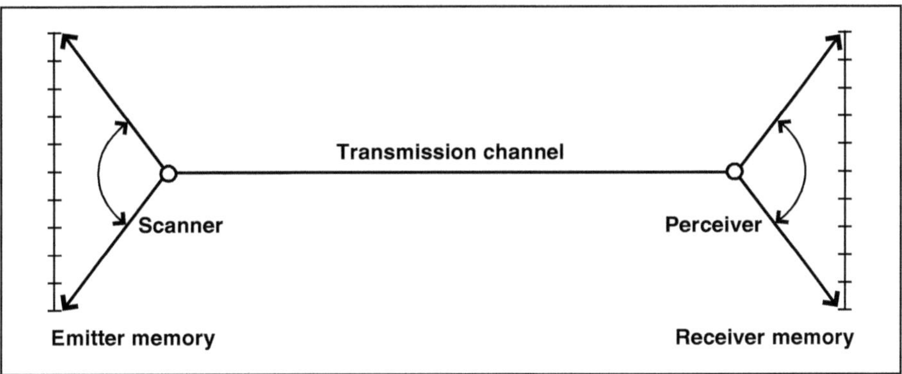

Fig. 5.1. Schematic representation of a communication system with emitter at left, communication link and receiver at right. Both have a memory which contains the message. In the receiver, the memory may not be changed by the message, because the question has not been asked.

and the transmitted entropy is defined as the entropy difference before and after reception.

$$\Delta S = S_{After} - S_{Before}$$

To give a numerical example, suppose that a memory element has 1024 states, each with a probability of 1/1024 before the message, that is an equal probability of each state. After reception the value of one probability is one (certainty) and all the others are zero. Hence

$$\Delta S = -(0 - 10) = 10$$

so the information gained is 10 bits.

Information can easily be lost in transmission if:

- the probability assessment is not changed by reception of the message, that is the message says what the receiver already knows. This is redundant information or redundancy;

- the message contains a signal that does not correspond to any element in the receiver's memory, meaning that the signal replies to a question that has not been asked, which is irrelevance or irrelevant information;

- the scanners become desynchronized;

- the signal is mutilated by noise.

A different concept is technical information as contained in drawings or computer programs, which can be read and transmitted by a computer. Many people including the authors have transmitted technical drawings over the Internet. The important point is that technical information can be copied and transmitted at will and is not erased by this process.

5.3 Micro-information and negentropy

The equality of the equations for entropy and information, apart from the factor k ln2, leads to the conclusion that statistical entropy and information are the same. So a flow of information and a negative flow of entropy, or negentropy, is the same.

The obstacle is that technical information has different conservation properties (it can be read many times) from macroscopic entropy.

The solution is to distinguish between micro-information and macro-information as follows

Micro-information refers to the individual DOF of matter, where each memory element corresponds to the state of memory;

Macro-information refers to macroscopically readable signals, where each signal value corresponds to many DOF or molecules.

Macro-information is therefore multiple redundant micro-information, and as such can be read and reproduced at will without being destroyed by the reading process.

In detail, micro-information is equal to statistical entropy, except for the factor k ln2, where k = Boltzmann's constant, with a dimension of J/K. This constant has been called the quantum of entropy, except that it is not fixed to integral values as is the other quantum constant h (Plank's constant), as mentioned above.

Micro-information is represented by the states of a DOF of an atom and if it is read, f. i. by a scanner, it is erased and the state of the memory after reading is indefinite.

Macro-information, on the other hand, can be read several times and is not destroyed, because it is multiple redundant micro-information, as in a technical memory. We can say that each state of the memory consists of many DOF of the order of Avogadro's number. Thus it has multiple redundancy, and the scanner disturbs only a small fraction of them: information can be read many times.

Here we also have the reason why Maxwell's demon cannot work: it sees only molecules which contain a few DOF, i.e. micro-information that is erased by reading.

5.4 Information theory, noise and organization

Transmission of information is impaired by noise in the electronic sense. Organization, in the biological sense, is a kind of ordering of parts of a system into a repetitive structure. Therefore an organized system possesses much redundancy. The biological aging process comes from the consumption of an initial redundancy in living cells.

Organization, in the biological sense, is a kind of order between parts of the system and is a compromise between maximum redundancy, as in a crystal, and maximum diversity, as in a gas. These ideas were developed in a philosophical book by Atlan in 1979, "Between Crystal and Smoke", original French "Entre le cristal et la fumée".

These ideas were also put forward in another book by Atlan (1972), where self-organization is likened to a change in organization towards increased efficiency under the effects of random environmental factors. He concludes with remarks on self-organization and the influence of temperature, which is needed to maintain the structure. In repetitive structures and in crystals, small doses of error or noise can be tolerated, whilst high doses destroy the system.

In metastable systems such as living organisms or organizations, small doses of noise may increase their functional organization, with an increase in complexity and a decrease in redundancy. This is what Atlan calls the "complexity principle" It indicates the possibility of the creation of information, which is missing in the original Shannon information theory.

In detail, the necessary conditions are:

- The existence of functional initial redundancy, which should be sufficient to prevent disorganization on a decrease in redundancy;
- The existence of at least two different levels of organization. This is because the usual decrease in a channel under the effects of noise becomes an increase in information content of a system.

No noise in the channel would mean that input and output are redundant from the viewpoint of the system. Examples are found in

1. The mechanisms of biological evolution induced by random mutations;

2. The maturation of an immune system with the creation of huge diversity;

3. The so-called epigenic instances of "developmental noise".

The notion of deterministic chaos also belongs to this area of thermodynamics and information theory. Here we should note that it is not sufficient to diagnose chaos just because we observe a system with nonperiodic oscillations. Rather, we must have advance knowledge of the dynamics of the system and, in most cases, chaos cannot be distinguished from periodic oscillations perturbed by random fluctuations. This follows also from the complexity principle (see above) with the role of superposition of repetitive order and random noise applied to it. For an easy introduction to chaos see [LURCAT 2002].

We hold this work at the frontier of biology and information theory as very significant and would like to see further research on it.

5.5 Applications

5.5.1 Brusselator and Prigogine's minimum entropy principle

As an example of coupled chemical reactions, we show one system that has been studied by Prigogine. It is called the Brusselator to commemorate the city where he worked.

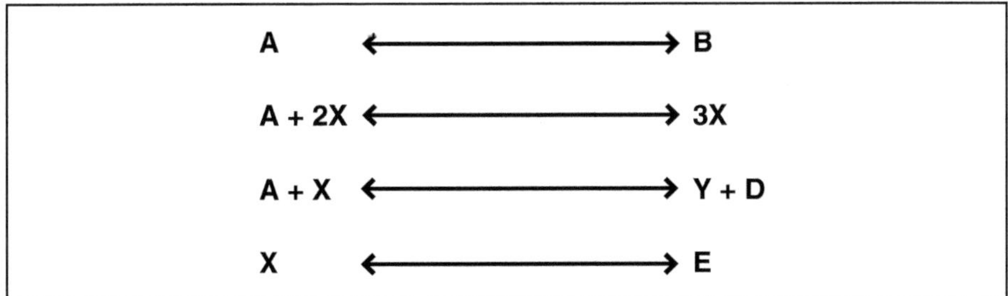

Fig. 5.2. Equations of the Brusselator, which is a system of coupled reactions.

Fig 5.2 shows the applicable equations. Note that what Prigogine calls irreversible thermodynamics is really thermodynamics that includes RS-multiports (a BG symbol), where all entropy is generated.

Reversible thermodynamics would then be a Bondgraph without RS-multiports, i.e. not operative, f.i. by low chemical tension on them[2]. This is very artificial and we prefer not to make the distinction between reversible and irreversible thermodynamics but to write a general BG instead. To explain further: for us, all elements of a BG are reversible except the R-elements, which are really power conserving, and entropy producing RS-multiports, and thus irreversible.

The Brusselator is a system of coupled reactions, written in chemical notation in Fig 5.2.

They can easily be represented by a Bondgraph such as in Fig 5.3 where some simplifications have been applied: all stripping and unstripping of the multiport-C and the sources are omitted, as well as all new entropy from the twoport-RS. These effects are still present and can give rise to chemical and thermal oscillations.

It is usually said that a system of reactions as in Figs 5.2 and 5.3 cannot oscillate near to chemical equilibrium, that is when the twoport-RS are linear in the affinities. On the contrary, far from equilibrium, when the equations are quadratic or higher, chemical oscillations can occur in a Bondgraph such as Fig 5.3. To prove this, a differential operation is performed on the chemical tension and the mass flow rate [PRIGOGINE 1967]. We are not convinced by this complex differential operation and say that oscillations can also come from one of the numerous neglected effects. The examples presented such as the prey-predator system are too simple to be convincing.

The difference between diffusion and convection cited on page 238 of [JI 1991] is, in Bondgraph terms, the difference between entropy conduction and entrainment by mass flow, treated in our section 3.5. He cites Prigogine who

---

[2] Or low current flowing through the resistor.

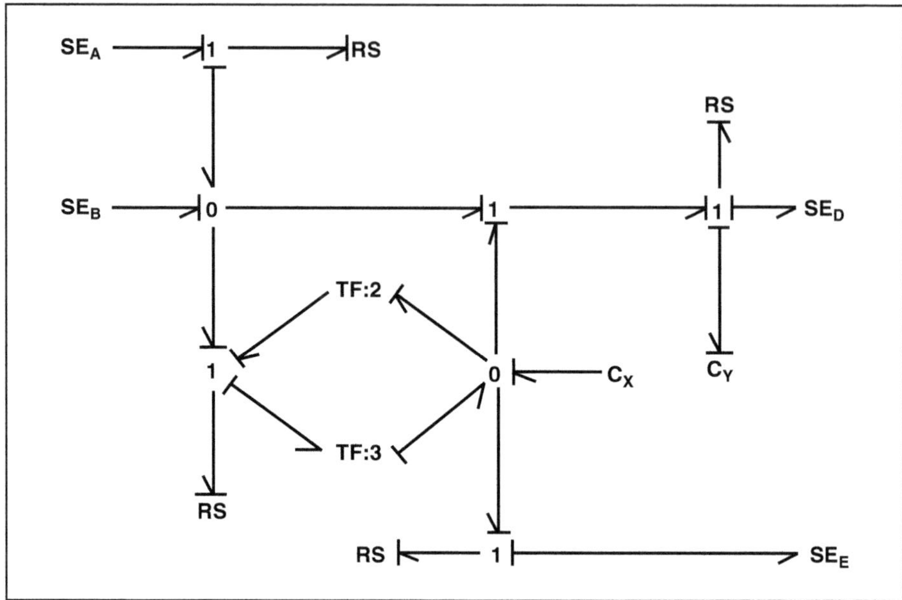

Fig. 5.3. Bondgraph for the Brussellator. The effort sources could by replaced by large capacitors.

says: convection of free energy feeds the living system, which we find is a nice formulation. For us, free energy is a notion only useful with constant temperature and we do not use it.

Here we also find Prigogine's principle of minimum entropy production, which is often cited [KONDEPUDI 1998]. It seems to us that this is a minimum dissipation theorem, originating from the old minimum dissipation in electric circuits [MAXWELL 1873], now over 100 years old. There, in a circuit of resistors driven by many effort (voltage) sources, the currents adjust themselves so that the overall dissipation becomes minimal. This would translate well into chemical networks represented by Bondgraphs. Dividing by the constant temperature, one has the minimum entropy produced.

If the circuit has resistors at different temperatures, minimum entropy production is no longer valid. As an example we take Fig 5.4 with many resistors where one resistor is taken out and brought to a lower temperature. One diminishes the resistor, itself a kind of potentiometer, such that its resistance remains the same in spite of the lower temperature, whereupon everything remains the same as far as dissipation is concerned. Entropy production is higher because one has to divide by a lower temperature.

To restate this conclusion, the resistors dissipate power and produce entropy flow. This means that the resistors adjust the currents so that, with a given setting of the voltage sources SE, dissipation is minimal. Resetting the SE would give another dissipation.

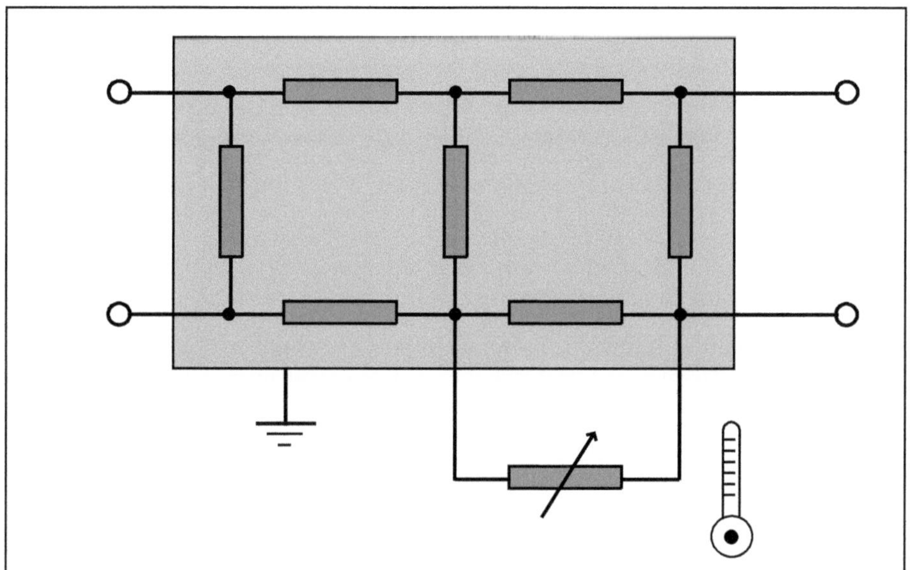

Fig. 5.4. Electric circuit with many resistors, one of which is outside the envelope. If this is a potentiometer and temperature is lowered, while resistance is kept constant, dissipation will remain the same, but entropy production is increased.

Next, one resistor is taken out and placed at lower temperature; the resistor can be adjusted (like a potentiometer) by an extra variable x. With lower temperature, the resistance can be set at the same value. Hence dissipation remains the same and Maxwells' mini dissipation is applicable, but entropy production is increased. So as a general conclusion we have minimum dissipation, and minimum entropy production is a misnomer.

There is also the interesting book by Ph. Ball (1999). He starts from the Brusselator and shows various oscillations, patterns and formation of Turing structures. Autocatalysis and inhibition are well explained. It is unclear whether the oscillations come from the RS-multiport far from equilibrium (Section 4.1) or from other neglected effects, such as entropy stripping. Perhaps the problem can be approached from the point of view of electronics with the conditions for producing oscillations in an electronic circuit. There we have an amplitude balance and a phase balance, which determine amplitude and frequency respectively.

5.5.2 Bhopalator

The Bhopalator as a model of living cells in Ji's book [JI 1991] is intriguing. Firstly, his figure 1.1 on page 16 can be expressed as a Bondgraph, with more or less detail. Secondly, his fig 1.7 on page 81 is a model of living cells. To us it seems that this pictorial model is really a complex Bondgraph, and an extension of the Brusselator (last section) and the chemical Bondgraphs in chap 4.1. Naturally, this figure is very complex and the Bondgraph will also be complex. But all the essential parts are there, including the dissipative structures of Prigogine, which would translate into coupled RS-multiports. In this sense, all living cells contain information, but, as said, the amount is small compared to thermodynamic information or entropy (section 5.1)

Also interesting is the relation to the big bang theory in Ji's book. It remains to be demonstrated how the postulate of information generation can be correlated with Shannon's entropy of a message (sec 5.2).

The transmission of information depends on an assessment of the probability of each state i before and after the message is received. So in the light of information theory, for each mechanical dimension, the information gained depends on the probabilities before and after receiving the message, which in turn depends on the reliability and ambiguity of a signal.

5.5.3 Information theory, diesel locomotives and capital

In sec 5.1 we have shown the notion of entropy of a message after Shannon with ambiguity, which amounts to a question that has not been put. This was in a sense the information content of a message. Here we want to extend the use of the concepts of information theory to capital and labor, as in economics.

Plants or physical capital can be characterized by their information content. Basically we are dealing with mechanical engineering, but the reasoning can be applied to computers and probably extended to software.

A design is specified by its general arrangement drawings and many detail drawings. There are many dimensions or lengths, and each dimension has a certain tolerance. For example, a length may be 100 mm and have a tolerance of $+/-0.01$ mm. The information content of this dimension (length) is

$$I_{Each\ Dim} = \log_{\text{Dual}}\left(\frac{L}{\Delta L}\right) \tag{5.7}$$

where L = length and ΔL = tolerance. This has to be summed for the dimensions of all the different parts of the machine to arrive at the total information content of the machine

$$I_{Cap} = N \sum \log_{\text{Dual}}\left(\frac{L}{\Delta L}\right) \tag{5.8}$$

N = number of different machine parts.

Note that if a machine comprises many equal parts, e.g. screws, their info capital is only taken once.

This is only a first approximation, which takes no account of the cost of materials and the rapid increase in cost with machine size.

Box: Steam and Diesel Locomotives
We estimate, very roughly, the information content of the locomotive specification to be:

$$I = N \ \log \ d \ (q)$$

$$q = \frac{L}{\Delta L}$$

$$N = Number \ of \ specs$$

$$L = Length$$

$$\Delta L = Tolerance$$

Steam Loco $N = 8000, q = 1000 \text{ mm}/0.1 \text{ mm} = 1.0\text{E}3$
$I = 8000 \ \text{logd}(1000) = 80\text{E}3 \text{ bit} = 80 \text{ kbit}$

Diesel Loco $N = 30000, \ q = 100 \text{ mm}/0.05 \text{ mm} = 2.0\text{E}3$
$I = 30000 \ \text{logd}(2000) = 330\text{E}3 \text{ bit} = 330 \text{ kbit}$

Energy Consumption

Steam $\dot{E} = 25$ MWthermic ($= 3.3$ tons coal/hour) $I = 80$ kbit

Diesel $\dot{E} = 8$ MWthermic ($= 360$ kg oil/hour) $I = 330$ kbit

Difference $\Delta\dot{E} = 20$ MW $\Delta I = 250$ kbit

So more information in the Diesel loco makes for less energy consumption.

Over a machine life of 10 years ($= 320\text{E}6$ sec), this makes $\Delta E = 20$ MW*320E6 sec, or 6.4E15 J, which allows us to define a macroscopical quantum of energy Eq as follows

$$Eq = E/I = 6.4\text{E}15 \text{ J}/250\text{E}3 \text{ bit} = 26\text{E}9 \text{ J}$$

Here one takes information as dimensionless, in spite of it being entropy related.

Transmission of Technical Information

The output of human labor can also be conceived as an information stream. Like electronic signals, it is subject to random noise, represented by careless mistakes on jobs, which depend on the exercise of care and attention. The information stream from a human being has been estimated at between 40 and 1000 bits/sec.

In manufacturing, specifications and standards are used, which are part of the knowledge base of each firm. They also provide information, and there is yet more information in the knowledge of the personnel. Technology transfer agreements between firms are really a rapid form of information transmission. This again has connections to deliberately false messages and cryptography.

In the box we have estimated the infocap, or information content of capital for the older steam and the newer diesel locomotives. We relate this to the energy or fuel consumption of the locomotives and arrive at a theoretical amount or quantum of energy of E = 20E5 J. These points should be developed further.

While we are dealing with speculations, there is also the concept of gnergy by Ji 1991, a mixture of energy and information. It is

$$g_n = I + E \tag{5.9}$$

I = information, related to entropy, E = total energy, in fluids normally expressed as enthalpy H or enthalpy stream $\dot{H}$. What, we wonder, would be the connection with chemical tension (potential), which has the formula

$$\mu = \frac{I}{n} + \frac{T\,S}{n} \tag{5.10}$$

It appears, then, that Ji's "gnergy" is really another expression for chemical tension. Anyway, the book is well worth reading and thinking about, especially his fig 1.15 (page 156) on the evolution of the universe, where energy becomes less and information increases over the billions of years due to the appearance of life.

Finally, his section 1.4.2 with the classification of machines and his fig 1.1 (page 15) would become much clearer if expressed by Bondgraphs. After all, Bondgraphs have been used in biology for 30 years [OPK 1973].

5.5.4 Solar energy and thermal balance of the earth

Since solar energy and global heating have been the subject of much discussion from 1970 to the present day (2006), we treat it here as part of thermal radiation. In fact solar radiation is a so-called black body radiation which was studied by Max Plank around 1880 as part of his doctoral thesis. Later, in 1900, he established the laws of quantum mechanics. It is radiation with a temperature of 6 000 K and, in respect of the part intercepted by the Earth, a power of 175 E15 Watts.

On the Earth, many processes and all living things consume chemical tension and can store entropy by the mechanism of entropy stripping. So all animal matter consumes chemical tension, that is chemical potential which is taken in as food. This is burned down to carbon dioxide and entropy. Plants are different: they produce oxygen and foodstuffs, driven by the light from the sun.

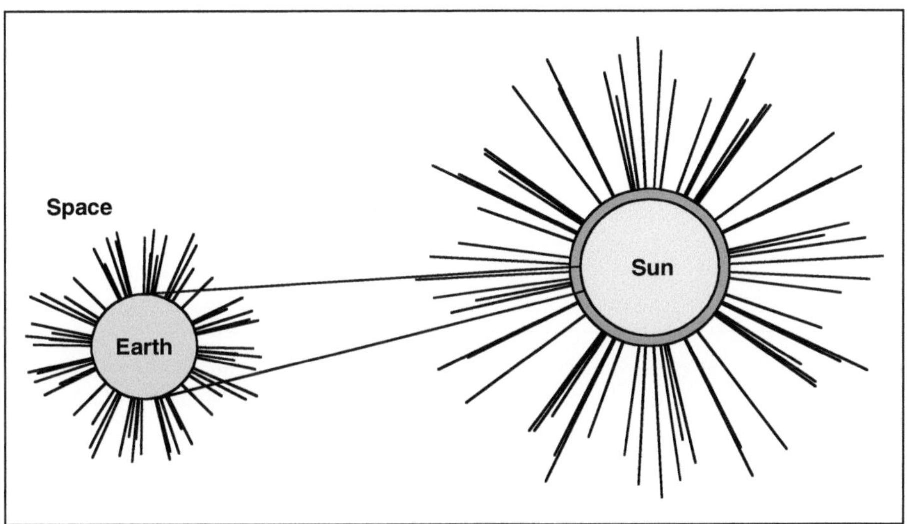

Fig. 5.5. Solar radiation as coming out from a cavity through a small hole, the said black body radiation, with a temperature of 6 000 K of which 155 PeW (175 E15 Watts) hit the earth. This entrains an entropy stream of 39 TeW (39 E12 Watts).

The temperature of the sun depends somewhat on the method of measurement, see Fuchs 1996 . This is shown schematically in Fig 5.5. The entropy of black body radiation is higher by a factor $1.33 = 4/3$ [THOMA 2000]. This is actually one of the properties of black body radiation, as shown by Plank about 1880. It was one of the foundations of quantum mechanics. This radiation is sometimes referred to as "photon gas".

Hence one can produce an energy and entropy balance of the earth. Sunlight brings a power of

$$\dot{E} = 175 \; P \; W = 175 \; E \; 15W \tag{5.11}$$

with a temperature of 6 000 K. The corresponding entropy flow is

$$\dot{S} = \frac{4\dot{E}}{3T} = 39\frac{T \; W}{K} \tag{5.12}$$

We have here a factor of $4/3 = 133\,\%$ which comes from the special properties of thermal radiation [THOMA 2000]. So our Carnot equation in section 1.1 must be modified by this factor, which is not far from one. Hence all our qualitative considerations remain valid.

The same amount of power is radiated back into space as shown in Fig 5.6. This takes place mostly in high clouds where the temperature is lower, about 250 K. The consequent entropy flow is

$$\dot{S} = \frac{4\dot{E}}{3T} = 933\frac{T \; W}{K} \tag{5.13}$$

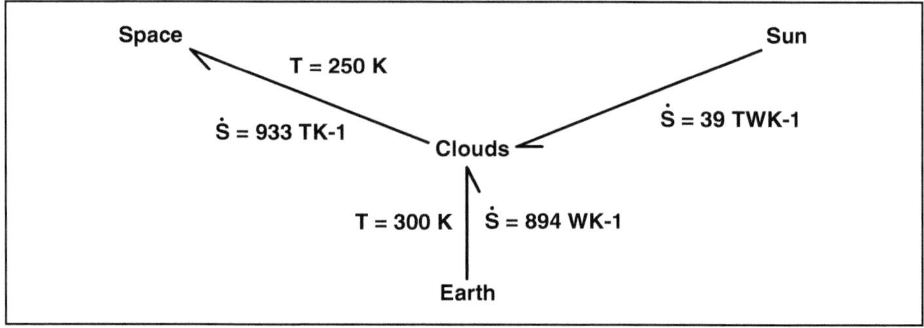

Fig. 5.6. Schematic representation, a quasi Bondgraph, of the power and entropy flows of the earth. The difference is large enough to absorb all entropy that is produced by humans.

Climate Change and Global Warming

As an excursus, one must always take into account the accuracy of technical data or information which can be roughly classified as follows:

1. Order of magnitude accuracy, as in our data about the earth
2. Engineering accuracy to about 1% = 10.0E-3, which is required for stress calculation for instance
3. Physical standard accuracy, which should be 10.0E-6 or better

The topic of climate change and global warming is fashionable today, but it has been around at least since the time of Clausius, ca 1860. At that time they spoke about the thermal death of the earth, following from the concept of entropy. The argument was that all processes are connected with entropy generation, therefore the earth will ultimately be flooded with it and become uninhabitable: thermal death after Clausius.

This argument is independent of the notion of entropy: one could argue that all processes have some friction and therefore some dissipation. So they produce heat and over time the earth becomes too hot to live on. This is not so, because only excess heat is radiated back into space. The atmosphere is approximated as a black body and an increase of one mK (Millikelvin) is enough to radiate away all entropy flow from human activity. Anyway, this flow is negligible compared to entropy generation in nature by the decay of dead plants and animals. So, as long as the sun remains at 6 000 K and space at 4 K, there is nothing to worry about, at least for the next 100 million years.

About the CO_2 balance of the earth, we have an article [NZZ 1990] that gave a carbon dioxide flow of 33E3 kg/sec from total human activity compared to a natural flow of 3.3E6 kg/sec from forests and 3.0E6 kg/sec from plankton in the oceans. The additional flow of entropy due to humans can easily be taken away by a minute increase in the temperature of the clouds and a little more radiation into space.

So news about impending thermal death is unwarranted and is perhaps something of an invention by scientists seeking funds for research and travel. There have been climate changes in geological time, say about 200 million years ago, and there were ice ages up to 20 thousand years ago, but these were not due to human activity. For comparison, dinosaurs were spread throughout the Earth until they died out quite suddenly about 63 millions years ago. In this time scale, humans have existed for about 6 million years.

Negentropic city

The term negentropy simply means negative entropy and was introduced by E. Schrödinger in his book "What is Life". This little book was very influential.

Pneumatic cars are an alternative to electric ones and also need no air for combustion. There is however the problem of refueling which was calculated in Figs 3.11 and 3.12. They work with air at high pressure, say 20 Mpa, and at room temperature. The difficulty is to design an air engine for such high pressures and room temperature, and the associated heat exchanger that takes entropy from the environment. Insofar as air can be taken as an ideal gas, its internal energy depends only on temperature, and not on pressure. So, putting air under pressure means that compressed air has a deficit of entropy, that is to say too much negentropy in the sense of Schrödinger, which has inspired in some thinkers to imagine a negentropic city like shown on Fig 5.7.

The cars are supplied with negentropy, which is nothing other than compressed air, and take entropy from the environment. This negentropy is taken from the ocean which is warm on the surface and cold deep down as well as from the earth which is cold on the surface and warm deep down.

We discussed this with Cesare Marchetti and discovered that pneumatic cars have better endurance than electric ones, but cannot compete with small highly developed gasoline engines. Also, air at these high pressures and room temperature ceases to behave as an ideal gas and therefore these considerations only have approximate validity. Nevertheless we found the idea attractive and have included it from [THOMA 1977].

5.5.5 Philosophical questions

Our thermodynamics is a return to Sadi Carnot (1796–1832) and the theory of caloricum as heat which was fashionable in his time.

Where we differ is that entropy, as caloricum is called today, enters all bodies in order to heat them. Only entropy is not conserved, as caloricum was, but

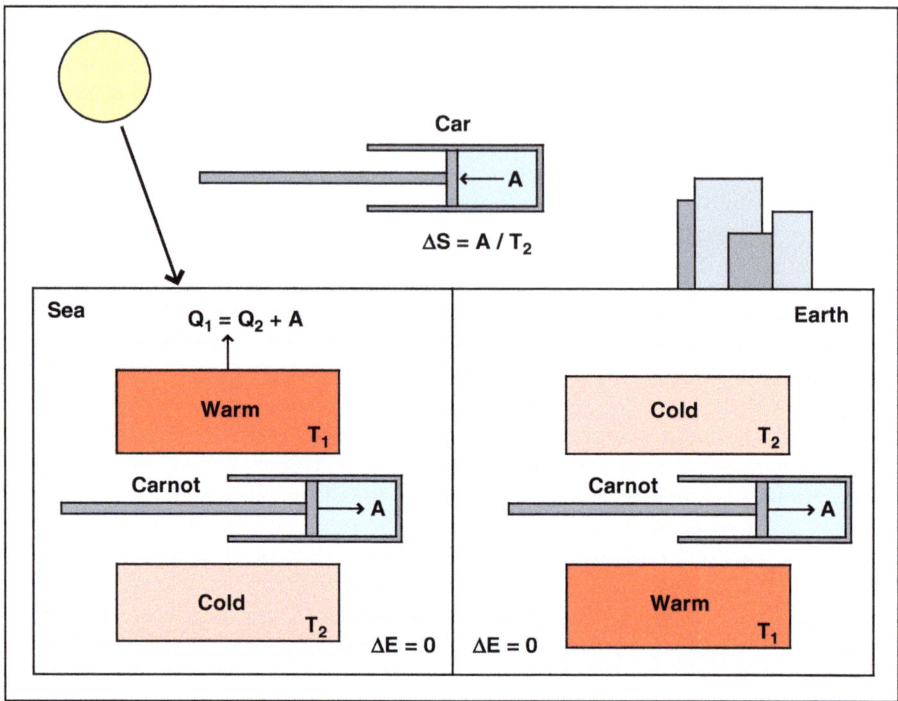

Fig. 5.7. Schematic of a negentropic city with power coming from the sea and from the earth. It comprises some Carnot cycles as indicated by warm and cold reservoirs and some pistons and cylinders.

semiconserved: it can never be destroyed but can be generated by all sorts of frictions. In a sense, entropy generation gives us the direction of time.

In our Bondgraph approach, flows like entropy flow always go from higher to lower effort and never the other way. This is different from particle dynamics and particle flows, where each flow is accompanied by a reverse particle flow, the principle of detailed balancing. Macroscopic effort gives a very small pre-ferred flow component to each particle, f. i. the movement of electrons. This gives the macroscopic flow, i. e. the electric current.

All theories linking microscopic to macroscopic phenomena are dubious, f. i. the Eyring theory of chemical kinetics and their temperature dependency. They should be examined in the light of thermodynamic fluctuations, which provides a link between macrophysics and microphysics.

We can only say that macrophysics is a worthwhile and interesting endeavor, which can lead to the design of machines. Any excursion to microphysics should be well documented from macro-physical facts.

Bibliography

ATLAN 1972

L'organisation biologique et la théorie de l'information
H. Atlan
Hermann, Paris, 1972, to be reprinted by Editions Seuil, Paris, 2006
Reprinted in English as chapter 1 in
Cybernetics, Theory and Applications
Ed. R. Trappl
Springer, 1982
"This early work is important to show the significance of information."

ATLAN 1973-1

Resistance and Inductance-like Effects in Chemical Reactions, Influence of Time Delays
H. Atlan and G.Weisbuch
Journal of Chemistry, 11, pp. 479 – 488, 1973
"Speculative example of chemical inductance."

ATLAN 1973-2

Tellegen's Theorem for Bondgraphs, its Relevance to Chemical Networks
H. Atlan and A. Katchalsky
Currents in Modern Biology, North Holland, Vol 5, pp 55 – 65, 1973

ATLAN 1974

On the formal definition of organization
H. Atlan
Journal of Theoretical Biology, 45, pp 295-304, 1974

BALL 1999

The Self-made Tapestry
P. Ball
Oxford University Press, Oxford, 1999

CALLENDAR 1911

Presidential Address
H. L. Callendar
Proc. Phys. Soc., pp. 153-189, London 1911
"Historically important, because it shows the first use of entropy as thermal charge after Carnot."

DENBIGH 1971 — **The Principles of Chemical Equilibrium**
K. Denbigh
Cambridge University Press, Cambridge, 1971

DUBBEL 1986 — **Taschenbuch für Maschinenbau**
Dubbel
15th Edition, Springer Verlag, Heidelberg, 1986
"Large comprehensive german handbook of engineering practice with many drawings."

FALK 1968 — **Theoretische Physik II - Allgemeine Dynamik, Thermodynamik**
G. Falk
Springer, Heidelberg, 1968

FALK 1976 — **Energie und Entropie**
G. Falk and W. Rupper
Springer, Heidelberg, 1976

FALK 1977 — **Konzepte eines zeigemässen Physik Unterichts**
G. Falk and H. Hermann
H. Schroedel Verlag, Hannover, 1977 to 1982
"Series of brochures made for secondary education in physics, reflecting the thoughts of Falk. Well worth reading. Job 1981 is in them."

FALK 1985 — **Entropy, a Resurrection of Caloric – A look at the History of Thermodynamics**
G. Falk
European Journal of Physics, pp 108-115, 1985

FUCHS 1996 — **The Dynamics of Heat**
H. Fuchs
Springer, Heidelberg, 1996
"Treatment of thermodynamics without BG, but including chemical tensions with temperature and pressure dependency."

GDT 2000 — **Les Bond Graphs**
G. Dauphin-Tangy, Ed.
Hermes, Paris, 2000
"Treatment of BG by several authors including J. Thoma. Here, the notion of causal path appears and also the element SW or SWITCH appears, simple for an electric circuit but difficult for a BG."

JI 1991 — **Molecular Theories of Cell Life and Death**
S. Ji
Rutgers University Press, 1991
"Interesting and intriguing."

JOB 1972   **Die Entropie als Wärme**
G. Job
Academische Verlagsgesellschaft, Frankfurt, 1972
"Early publication on thermodynamics with entropy as thermal charge."

JOB 1981   **Die Werte des chemischen Potentials**
G. Job
Schroedel Verlag, Hannover, 1981
"Interesting treatment of chemical tension as a basic variable of chemical reactions, with its pressure and temperature dependency, republished by Fuchs 1996 in his section 4.2.1 and table 13."

KARNOPP 1979   **State Variables and Pseudo-BG for Compressible Thermofluid Systems**
D. C. Karnopp
Transactions ASME, JDSMC, vol 107, pp 241-245, 1979
"Basic formulation of vector BG with pressure and temperature as efforts and mass flow and enthalpy flux as flows."

KMR 2000   **System Dynamics - Modeling and Simulation of Mechatronic Systems**
D. C. Karnopp, D. L. Margolis and R. Rosenberg
Wiley, New-York, 2000
"New edition of a famous textbook."

KONDEPUDI 1998   **Modern Thermodynamics – From Heat Engines to Dissipative Structures**
D. Kondepudi and I. Prigogine
Wiley, New York, 1998
"A good book on conventional thermodynamics."

LURCAT 2002   **Le chaos**
F. Lurçat
Collection que sais-je?, Presses Universitaires de Fance, Paris, 2002
"Excellent short introduction with reflections on the work of Poincaré."

MAXWELL 1873   **On the Theory of Electrified Conductors, and other Physical Theories involving Quadratic Functions**
J. Maxwell
Proceedings of the London Mathematical Society
Vol 22, pp 42-44, 1873
"Minimal dissipation theorem."

98   Bibliography

MENDOZA 1960   **Reflections on the motive power of fire – By Sadi Carnot -1824**
E. Mendoza
Dover, New York, 1960
"Classic with good remarks from Mendoza."

MOCELLIN 2001   **The Bond Graph Method applied to Social and Life Sciences**
G. Mocellin
Proceedings of the International Conference on Bond Graph Modeling and Simulation
2001, Vol. 33, Number 1, pp. 390-397

NZZ 1990   **CO_2 balance of nature**
Neue Zurcher Zeitung, 1990-10-27
"A daily newspaper based in Zurich, Switzerland."

OPK 1973   **Network Thermodynamics: Dynamic Modeling of Biophysical Systems**
G. Oster, A. Perelson and A. Katchalsky
Quarterly Reviews of Biophysics, Nr 6, pp 1–134, 1973
"Old publication with chemical tension and biophysics."

PIGUET 2001   **CampG/Sysquake, an Integrated Environment to Understand Dynamic Systems and Design Controllers**
Y. Piguet, J. Granda and G. Mocellin
Proceedings of the International Conference on Bond Graph Modeling and Simulation
2001, Vol. 33, Number 1, pp. 158-163

PIPPARD 1966   **Elements of Classical Thermodynamics – For Advanced Students of Physics**
A Pippard
Cambridge University Press, Cambridge, 1966
"Good and clear treatment."

PLANK 1958   **Physikalische Abhandlungen und Vorträge**
M. Plank
Volume 3, p. 261
Vieweg, Wiesbaden, 1958
See also his paper: "Zur Geschichte der Auffindung des physicalischen Wirkungsquantums"
"Collected Works, 1947, which is still well worth reading in 2006.
Classical papers in which he writes that he had success with black body radiation because he concentrated on entropy, not on heat."

PRIGOGINE 1967   **Introduction to the Thermodynamics of Irreversible Processes**
I. Prigogine
Wiley, New York, 1967
"Interesting old publication on entropy generation."

SOUTIF 2002   **Naissance de la Physique de la Sicile à la Chine**
M. Soutif
EDP Sciences, Les Ulis, 2002
"Intresting French book on the history of physics - in Asia and Europe."

THOMA 1971   **Bondgraphs for Thermal Energy Transport and Entropy Flow**
J. Thoma
Journal of the Franklin Institute, Vol 292, pp. 109-120, 1971
"First publication by Prof. J. Thoma on entropy generation."

THOMA 1975   **Introduction to Bond Graphs and their Applications**
J. Thoma
Pergamon Press, Oxford, 1975

THOMA 1977-1   **Energy, Entropy and Information**
J. Thoma
IIASA memo RM-77-32, 27 pages, 1977
"Bringing entropy as thermal charge and statistical entropy with information theory together."

THOMA 1977-2   **Network Thermodynamics with Entropy Stripping**
J. Thoma and H. Atlan
Journal of the Franklin Institute, Vol. 303, No 4, pp. 319-328, 1977
"Description of entropy stripping."

THOMA 1978   **Entropy Radiation and Negentropy Accumulation with Photocells, Chemical Reaction and Plant Growth**
IIASA memo RM 78-14, 24 pages, 1978
"Applications of entropy and black body radiation."

THOMA 1985   **Osmosis and Hydraulics by Network Thermodynamics and Bond Graphs**
J. Thoma and H. Atlan
Journal of the Franklin Institute, Vol.319. pp 217-226, 1985
"Discussion of osmosis."

THOMA 1990 **Simulation by Bondgraphs**
J. Thoma
Springer, Heidelberg, 1990
"Based on Prof. Thoma graduate course at the University of Waterloo."

THOMA 2000 **Modeling and Simulation in Thermal and Chemical Engineering**
A Bondbgraph Approach
J. Thoma and B. Ould Bouamama
Springer, Heidelberg, 2000
"Boilers and chemical engineering."

WIBERG 1972 **Die chemishe Affinitaet**
E. Wiberg
Walter de Gruyter, Berlin, 1972
"Entropy capacity compared to a wine glass and chemical tension."

Appendix 1

Understanding with Bond Graphs

Bondgraphs are interdisciplinary and consist of elements with some lines called bonds in between, just as we find in electronic circuits. When he invented them, Paynter was impressed by their (superficial) resemblance to chemical bonds, hence the name.

The lines carry special signs and the elements are letter codes from electrical engineering. Basically Bondgraphs relate the notion of voltage and current to all scientific definitions, called in Bondgraphing efforts and flows.

The power equation is always

$$\dot{\mathbf{E}} = \mathbf{e}\,\mathbf{f}$$

or power equals effort times flow.

The above equation is valid for true BGs. In other words, there also exists the so-called pseudo-BG where the product of effort and flow is not a power.

A1.1 Elements

In thermal engineering we have two kinds of pseudo BGs, namely

1. A BG with temperature as effort, and heat flux - not entropy flow - as flow variable. This is useful for problems of friction, which include heat conduction. There, heat flux is conserved, not entropy flow (Sec 2.4).

2. In pipes with hot gases or more generally with a moving mass, we take as efforts the pressure and temperature, and as flows mass flow and enthalpy flux, as explained in section 3.8. Thus we have a pseudo-vector-BG.

Essential to the concept are the number of wires or ports on each element, which also allow classification into oneport, twoport etc. elements. The ports connect to the other elements and the environment:

1. Causality can be inverted on C- and I-elements, but this then leads to derivative causality, which the computer does not like;

2. The old convention for the half-arrow at MIT was effort on top, flow below. Newer and better is flow on the side of the half-arrow, effort on the other side[1];

3. Sign conventions: power-in with C-, I-, and R-elements (which makes them stable when positive) and power-out on sources;

4. Signs are free in junctions, but the plus or minus signs are given on the corresponding block diagram.

In order to comply with both the MIT and the newer standard, we tend always to write the half arrow on the lower side of horizontal bonds, because this habit has propagated.

A1.1.1 Bonds and connections

These elements are:

Bondgraph symbol	Description	Equation
e f	**Simple bond:** Transmission of power, time derivative of energy, as a product of two factors: effort e and flow f.	$\dot{E} = e\,f$
e f	**Power flux positive:** Half arrow giving the positive sign convention.	
e f	**Direct causality**[2]: Effort to the right; Flow to the left.	

---

[1] Sometimes it is useful to know what is effort and what is flow...

[2] Causality, direct or inverse, is used in a different sense in electrical circuits, namely that the response of a filter cannot exist before the excitation; both meanings of causality have no relation.

Bondgraph symbol	Description	Equation
	Inverse causality[3]: Effort to the left; Flow to the right.	
	Signal or activated bond: Connection in which one of the pair of variables is neglected.	

A1.1.2 One port elements or one ports

These elements are:

Bondgraph symbol	Description	Equation
	C-element or capacitor: Integrates flow.	$e = \frac{1}{C} \int f \, d\,t$
	I-element or inductance: Integrates effort.	$f = \frac{1}{I} \int e \, d\,t$
	R-element or resistor: Dissipator or power sink with free causality.	$e = K\,f$ $f = \frac{1}{R} e$
	Effort source: Supplies or withdraws power.	$e = \text{constant}$
	Flow source: Supplies or withdraws power.	$f = \text{constant}$

We apply to all symbols the preferred integral causality.

For the sources, the causalities are compulsory, as for the detectors.

From an SE an effort or piston is pushing, from an SF, a flow or needle is pointing.

---

[3] One says, for short, effort is like a piston that pushes, flow like a needle that points.

A1.1.3 One and a half ports

These elements connect Bond Graphs and Block Diagrams, so they have one bond and one connection. Therefore we call them one-and-a-half ports.

These elements are:

Bondgraph symbol	Description	Equation
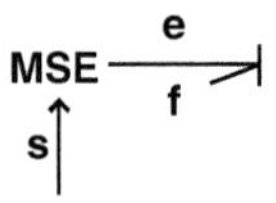 MSE $\xrightarrow{\ e\ }{}$ f, s	**Modulated source of effort:** Source modulated by a control signal s.	$e = K\,(s)$
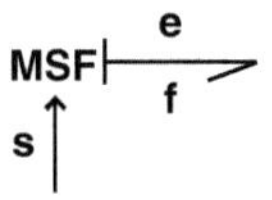 MSF $\xrightarrow{\ e\ }{}$ f, s	**Modulated source of flow:** Source modulated by a control signal s.	$f = K\,(s)$
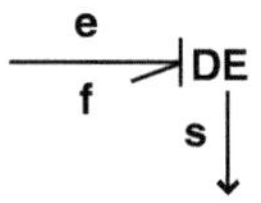 $\xrightarrow{\ e\ }$ DE, f, s	**Detector of effort:** Produces a signal s for a controller.	$s = K\,e$
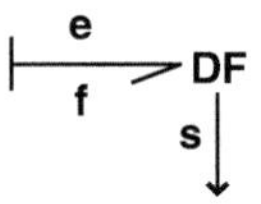 $\xrightarrow{\ e\ }$ DF, f, s	**Detector of flow:** Produces a signal s for a controller.	$s = K\,f$

K is a constant gain of sources or detectors.

With the modulated sources, our terninology "push and point" is especially useful: from a MSE an effort or piston is pushing, from a MSF a flow or needle is pointing. This helps to check causalities of BGs, especially of junctions.

A1.1.4 Two port elements or two ports

These elements are :

Bondgraph symbol	Description	Equation
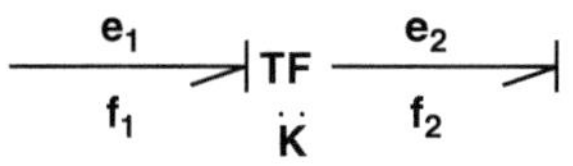 e_1, f_1 $\xrightarrow{}$ TF $\xrightarrow{}$ e_2, f_2, K	**Transformer:** Transmits power in the same or in another energy domain.	$e_2 = K\,e_1$ $f_1 = K\,f_2$
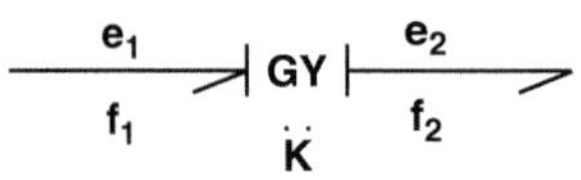 e_1, f_1 $\xrightarrow{}$ GY $\xrightarrow{}$ e_2, f_2, K	**Gyrator:** Transmits power in the same or in another energy domain.	$f_2 = K\,e_1$ $f_1 = K\,e_2$

Bondgraph symbol	Description	Equation
e_1 ⟶ MTF ⟶ e_2; f_1, f_2; s	**Modulated transformer:** Transformer modulated by a signal s from a controller.	$e_2 = K(s)\, e_1$ $f_1 = K(s)\, f_2$
e_1 ⟶ MGY ⟶ e_2; f_2, f_2; s	**Modulated gyrator:** Gyrator modulated by a signal s from a controller.	$f_2 = K(s)\, e_1$ $f_1 = K(s)\, e_2$

K is a constant transformation modulus for transformers and gyrators.

K(s) is a function of the signal s.

A1.1.5 Junctions elements or three ports

Junctions can have more than three ports, but the classification of three ports is often convenient.

These elements are:

- The effort junction 0, parallel and power conserving junction on which all the flows are equals and;

- The flow junction 1, series and power conserving junction on which all the efforts are equals.

Bondgraph symbol	Description	Equation
e_1, f_1 ⟶ 0 ⟶ e_2, f_2; e_3, f_3	**Effort junction:** Receives an effort and distributes it among elements in which flow is equal.	$e_3 = e_1 + e_2$ $f_1 = f_2 = f_3$
e_1, f_1 ⟶ 1 ⟶ e_2, f_2; e_3, f_3	**Flow junction:** Receives a flow and distributes it among elements in which effort is equal.	$f_3 = f_1 + f_2$ $e_1 = e_2 = e_3$

There are two causality rules on junctions:

- In an effort junction (parallel or 0-junction), one effort is pushing and all the other efforts are leaving. Hence on such a junction all the flows except one are going away;

- In a flow junction (series or 1-junction), one flow is attacking and all the others flows are leaving. Thus on such a junction all efforts save one are going away.

We see that both junctions are dual to each other.

In the push/point terminology:

- In an effort junction (parallel or 0-junction), one bond pushes, all other bonds point;

- In a flow junction (series or 1-junction), one bond points, all others push.

A1.1.6 Multiport elements or multiports

These elements are :

Bondgraph symbol	Description	Equation
$\dashv \ \overset{e_1}{\underset{f_1}{\rule{0pt}{0pt}}} \ C \ \overset{e_2}{\underset{f_2}{\rule{0pt}{0pt}}} \ \vdash$	**Multiport capacitance:** Multi-energy storage with two flows as input.	$\begin{pmatrix} e_1 \\ f_2 \end{pmatrix} = f \begin{pmatrix} q_1 \\ q_2 \end{pmatrix}$
$\overset{e_1}{\underset{f_1}{\rule{0pt}{0pt}}} \ I \ \overset{e_2}{\underset{f_2}{\rule{0pt}{0pt}}}$	**Multiport inductance:** Multi-energy storage with two efforts as input.	$\begin{pmatrix} f_1 \\ f_2 \end{pmatrix} = f \begin{pmatrix} p_1 \\ p_2 \end{pmatrix}$
$\vdash \ \overset{e_1}{\underset{f_1}{\rule{0pt}{0pt}}} \ IC \ \overset{e_2}{\underset{f_2}{\rule{0pt}{0pt}}}$	**Combined I and C multiport:** Multi-energy storage with one flow and one effort as input.	$\begin{pmatrix} e_1 \\ f_2 \end{pmatrix} = f \begin{pmatrix} p_1 \\ q_2 \end{pmatrix}$
$\overset{e_1}{\underset{f_1}{\rule{0pt}{0pt}}} \ RS \ \overset{e_2}{\underset{f_2}{\rule{0pt}{0pt}}} \ \vdash$ $\dot{S}\,\vert\,T$	**Resistor source:** Multi-energy dissipation with entropy production.	$T\,\dot{S} = e_1\,f_1 - e_2\,f_2$

If thermal effects are disregarded, RS-multiports become simple R-elements;

It is often convenient to use a word-Bondgraph first, that is one with only words in it, to establish variables and causalities, later to be replaced by standard elements.

Our RECO, HEXA and TEFMA elements are really word-Bondgraphs, because they use these abbreviated words.

A1.2 Energy and power conservation

It is useful to distinguish energy conservation and power conservation as follows: power conservation is stricter and means that power is always conserved, such as in junctions and transformers or gyrators. Here therefore energy is also conserved.

Energy conservation alone means that energy can be momentarily absorbed by an element, changing its state. To return the element's state to rest, power must come out again. Hence energy is returned eventually. All C- and I-elements are of this kind: they are energy conserving but not power conserving and are also called storage elements.

In regard to the correct formulation for writing a BG that is universally readable, there is a french expression, "la cuisine informatique" or "the computer science kooking", coined by GTD (Madame Geneviève Dauphin Tangy of Lille, France), which sums up the idea that a BG should be usable with all computer programs. Only later, when a scientist goes to his programs, special tricks for each program can be used. Hence special computer science tricks from the 'kitchen' can be introduced at that stage.

As an example, Fig A1.1 shows a simple BG between a 0- and a 1-junction, where the bond leading to the transformer as been activated (an MIT expression to say that it is under control). This means that the corresponding flow is neglected. In other words, the corresponding feedback action is ignored. This is a perfectly admissible as Thoma learned at MIT.

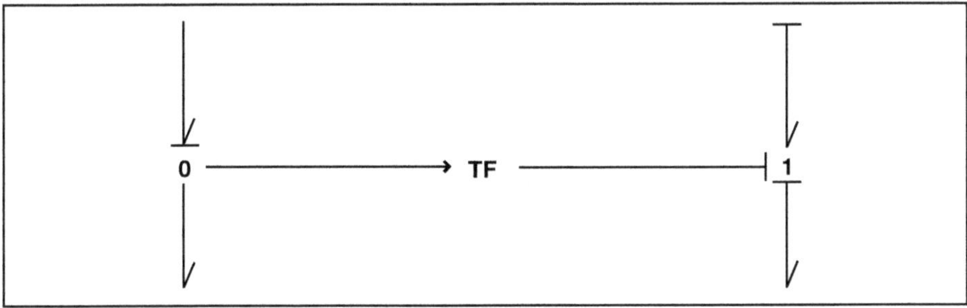

Fig. A1.1. BG of a transformer between an 0-junction and a 1-junction with one activated bond, as shown by the full arrow.

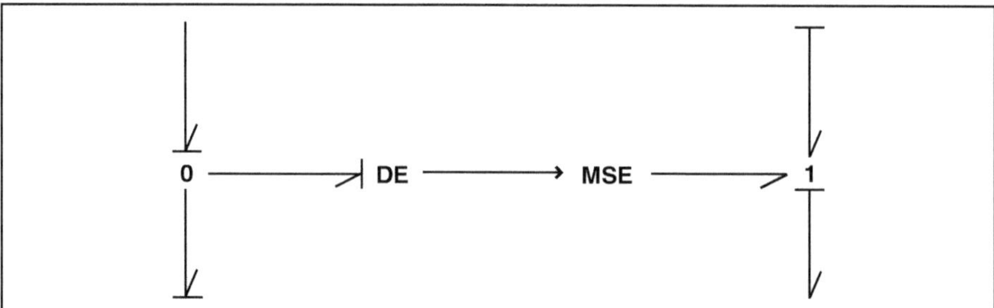

Fig. A1.2. A more correct BG with a detector of effort DE, a control signal connection and a modulated effort source.

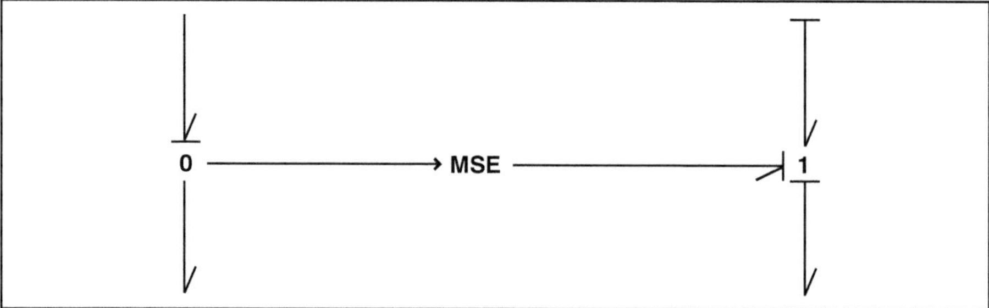

Fig. A1.3. BG with detector of effort suppressed, and with the modulated effort source directly driven by the zero junction.

Yet neglecting the flow means that the transformer becomes a simple gain, and would be better represented as such, as shown in Fig A1.2, with a detector of effort De and a modulated effort source MSE. So here we have a Block Diagram BD, between De and MSE. In this case the TF has turned into a gain K in a box, according to BD rules. It is not really necessary in this simple case, but very useful with large BGs.

In practice, with such a simple BG as Fig A1.2, one could omit the detector of effort DE and take the effort source directly into the modulated effort source MSE as shown in Fig A1.3.

So a signal taken from a 0-junction conveys the common effort in it, and from a 1-junction the common flow.

Fig A1.3 looks much better and more readable than Fig A1.1.

A1.3 Power signs

The normal orientation of power signs or half arrows should be as in Fig A1.4, namely into the oneports, out of the sources, and through the transformers and gyrators. This gives positive values for the parameters in almost all cases.

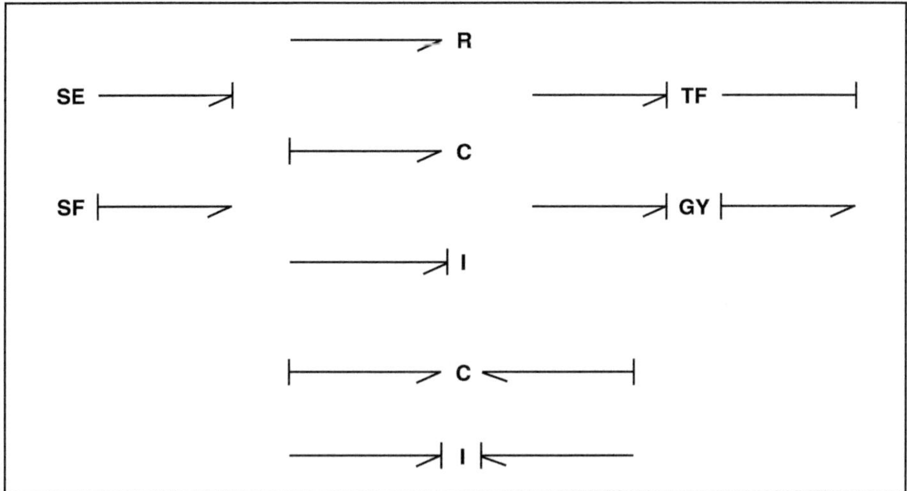

Fig. A1.4. Usual power signs for BG elements.

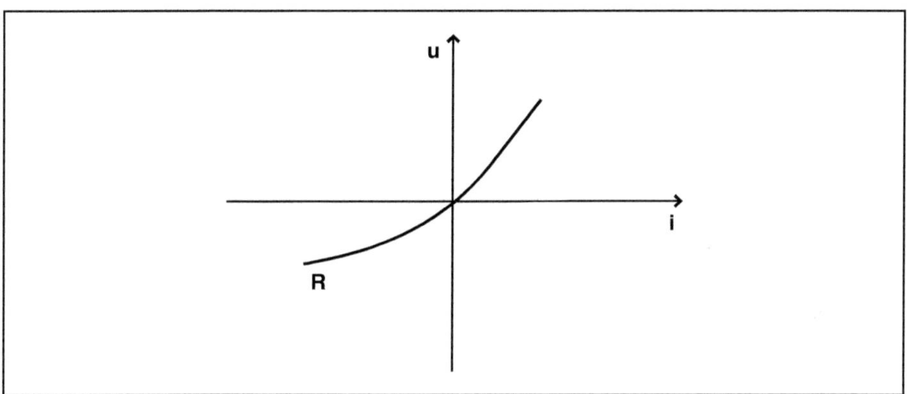

Fig. A1.5. With normal power orientation, resistors can be linear or nonlinear, but must be positive for positive effort (and negative for negative effort). Since power cannot come out, the quadrants 2 and 4 are forbidden.

Note that the half arrows for power only give the direction of flow of power, that is its positive direction. Thus effort and flow can both be positive or both negative, which fits many physical situations well.

An example is the R-element or its thermodynamic equivalent, the RS-field or RS-multiport. Here, as is well known from the electric case, power can flow into it but never out. So voltage and current can both be positive or both negative, but not one positive and one negative. This is shown in Fig A1.5 for a linear resistor. A P/N junction would be similar.

We can generalize this. A passive twoport does not deliver power to the circuit. Hence, the characteristic flow/effort curves can only go through the first and the third quadrants, whilst the second and fourth quadrants are prohibited by the second law of thermodynamics[4]. This is the irreversibility of entropy and thermal power flow in nature. Generally, all BG elements are reversible except the RS elements.

To repeat, the simple resistor and its BG-representation, the RS-multiport, is power conserving and irreversible. The causality on the thermal port is as shown, whilst the electric port is free from causality. Note that resistors with normal power orientation – inflow taken positive - are positive, although they can be linear or nonlinear.

In multiport R-fields or resistor networks, with two or more electric bonds and one thermal bond, the power condition is relaxed: only the thermal bond is irreversible, and the electric bonds can in part become negative. This is important in the so-called Peltier effect, that is the coupled flow of entropy and electricity.

A1.4 Negative resistances and negative C-elements

Occasionally negative resistances appear, in apparent contradiction to the second law of thermodynamics, as f. i. tunnel diodes in electronics, or simply negative springs, where we have less force with increased displacement.

These are thermodynamically possible over a limited range. In other words, the incremental resistance can be partly negative, on condition that total dissipation remains possible, as shown in Fig A1.6. It depicts a BG representation on top, where we have a flow source which produces a large amount of entropy, more than the entropy consumed by the negative resistance.

Also C-elements can have a negative part in their characteristics and have a similar characteristic to Fig A1.6 bottom. This is not contrary to the entropy (second) law of thermodynamics, but easily leads to instabilities.

A1.5 Compact units in pneumatics and hot gas

Compact units are very often used instead of pure S.I. units in oil hydraulics. The reason for this is that they give more impressive and easily remembered numerical values, which helps in the laboratory. An example is the displacement of hydrostatic machines, which is typically 20 to 200 ccm, more easily remembered than 20E-3 m^3. We have mentioned them in section 3.5 and they are fully described in Thoma 1990.

---

[4] If a measured device goes a little through the prohibited quadrants, it is evidence that it is not a pure R-element but has some C- or I-elements within it.

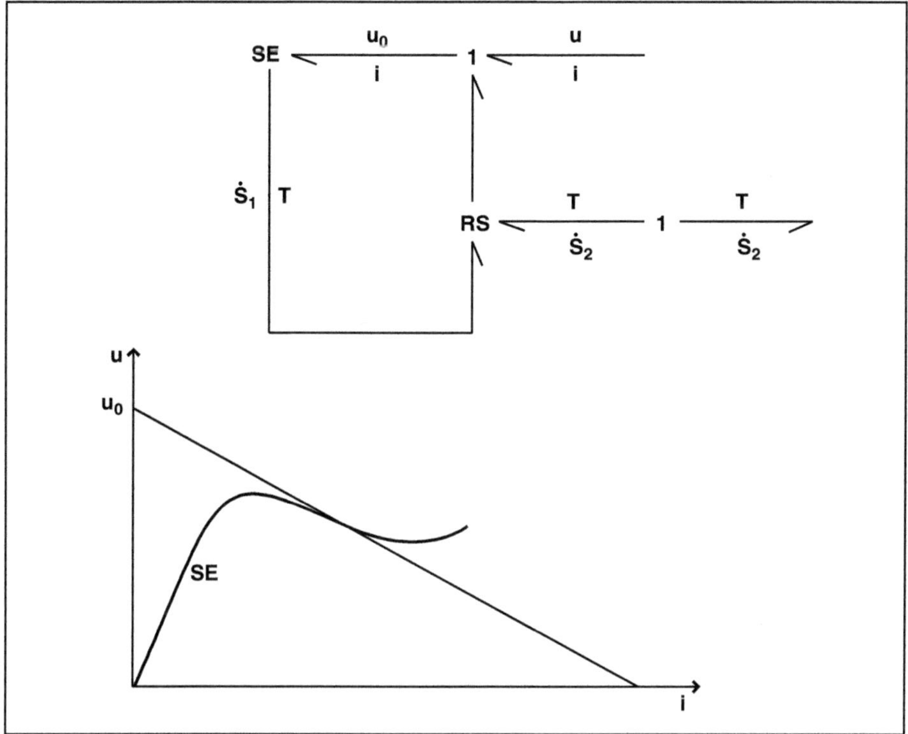

Fig. A1.6. Incremental resistances can be negative, as long as the total resistance remains positive. That means that more entropy is produced by the reverse driven flow source than absorbed by the negative resistance.

Less well known is the fact that, also in pneumatics and with hot gases, we can use different compact units because air is about 1000 times less dense than water. Indeed we use them in our vector BG for hot gases with thermal effects.

Pressure in [Mpa]

Mass in [kg]

Mass flow in [kg/s]

Temperature in [K]

Enthalpy flow in [kg/s]

Volume $1\mathrm{E}^{-3}$ m^3 = [lit]

Mass density $\rho = $ [kg/liter] = [kg/E^{-3} m^3]

Specific heat [kJ/kg K]

Example: atmospheric air at 0 Celsius:

$\rho = 1.29 \ [\mathrm{kg/m^3}] = 1.29 \ \mathrm{E}^{-3} \ [\mathrm{kg/liter}]$

$c_p = 2.005 \ [\mathrm{kJ/kgK}]$

Most inconvenient is the prefix kilo in kg, because it really means one thousand, a multiplier for units. Therefore we would like to go back to an old french unit

$$1 \ \text{grave} = 1 \, \text{kg, abbreviation } 1 \, \text{grv.}$$

This name was later changed in France to kilogramme [SOUTIF 2002]. We tend to use it because we have had computer programs fail as a result of kilo ambiguity.

A1.6 Multiport-C signs in thermodynamics and Maxwell relations

Maxwell reciprocity is a property of true BGs. Pseudo-BGs can also be used but then these interesting relations are lost. So there is nothing wrong in writing pseudo-BGs, and they are convenient for some heat conduction problems.

Maxwell relations belong to all multiport-C (and multiport-I) of BG and are expressed by

$$dU = e_1 \ d \ q_1 + e_2 \ d \ q_2 = U \ (q_1, \ q_2) \tag{A1.1}$$

In other words, from the formulation with the differential one concludes that the internal energy U is a function of q_1 and q_2 (and not of other variables).

By differentiating one obtains the corresponding efforts

$$e_1 = \frac{\partial U \ (q_1, \ q_2)}{\partial q_1}; e_2 = \frac{\partial U \ (q_1, \ q_2)}{\partial q_2} \tag{A1.2}$$

By deriving the cross derivatives once more

$$\frac{\partial U}{\partial q_1 \partial q_2} = \frac{\partial e_1}{\partial q_2} = \frac{\partial e_2}{\partial q_1} \tag{A1.3}$$

The equality of the cross derivatives (Schwartz's theorem) results in a relation between the derivatives of both efforts. This is the first of the Maxwell relations which now have minus in them. So they are also valid for the moving plates of capacitors, as we have shown in section 1.2. The minus signs result from the unfortunate choice of the internal energy in thermodynamics, which is expressed in flows.

$$\dot{U} = T\,\dot{S} - p\,\dot{V} \tag{A1.4}$$

In other words, volume flow is taken negative as the internal energy increases. Thus with gases and other substances, as volume decreases the compression energy increases. This choice in thermodynamics comes from the fact that volume itself is a positive quantity[5]. In other words, volume cannot become negative. However, no such restraint exists with volume flow, which can go in either direction and positive values can be assigned in either of them. So we could also assign positive volume flow to decreasing volume and thus avoid the minus sign in equation A1.4. This applies in principle, but in practice we keep to the convention, which states that a positive volume flow increases volume.

---

[5] This means that a negative volume cannot exist.

Appendix 2

Control Systems with Bond Graphs

BG originated as an attempt to write Block Diagrams for electro-hydraulic systems as a means of controlling them automatically, as we will show in the historical notes. Hence the relation between BG and BD is of interest, especially now that computer interactive environments exist to go from a BG to the design of a controller [PIGUET 2001].

We have seen that in a BG each bond carries a forward and a backward action, which is chosen with the causality: hence a complete, or in MIT terms, a fully augmented BG is equivalent to a BD, but usually impractically complex.

To bring it down to manageable size, engineering judgment is required to concentrate on the essential actions and to disregard the unimportant ones.

As an example, Fig A2.1 shows a control system with the dimensions of all variables added. It contains an input voltage which is mixed with the feedback voltage to give an error voltage. This is amplified into a current in an amplifier. Next comes the actuator which produces a force. In some hydraulic systems, this is a so-called electro-hydraulic servo valve followed by a cylinder. Then the load produces the speed. Each variable is characterized by a gain and the physical dimension of each signal is inscribed. The dimensions of each gain follow by dividing the output signal by the input signal.

The integrator is important as it calculates the position of the load. This goes into a pick-up or transducer (french capteur) that produces the return voltage, which goes back to the mixing point.

Two things should be noted on Fig A2.1 :

1. The transfer function of each element can have any dimension or unit, but the loop gain is dimensionless ;

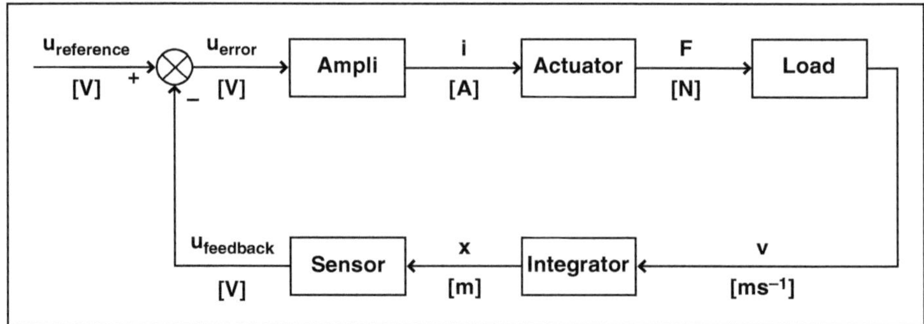

Fig. A2.1. Control system inspired by electro-hydraulics: at left the mixing point, where an error voltage is created, then the amplifier, actuator, load, integrator and sensor elements. The physical dimension of each signal is given in square brackets.

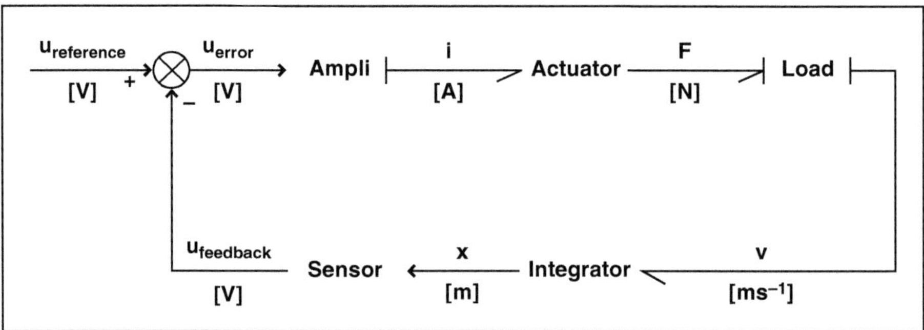

Fig. A2.2. Same control system by combined BD and BG with many activated bonds. Compared to Fig A2.1 it also shows the signs of the powers and uses the convention that the elements are not placed in boxes.

2. Normally in control systems the loop gain is greater than one and can become infinite for certain signal frequencies. This is the case in the example above, where the gain becomes infinite at low or zero frequency due to the integrator.

Fig A2.2 shows the same system as a BG with many bonds activated. This MIT parlance means that on these bonds one variable is neglected. In principle, all the activations would give secondary loops, and the designer must make sure that the gains are sufficiently small to be disregarded. As mentioned before, a BG is a systematic means of setting up a BD.

Proceeding now to automatic control engineering, one can simplify Fig A2.2 into Fig A2.3 with one forward gain F and a return gain G. These gains can be time dependent or a function of the signal frequency. Usually they are

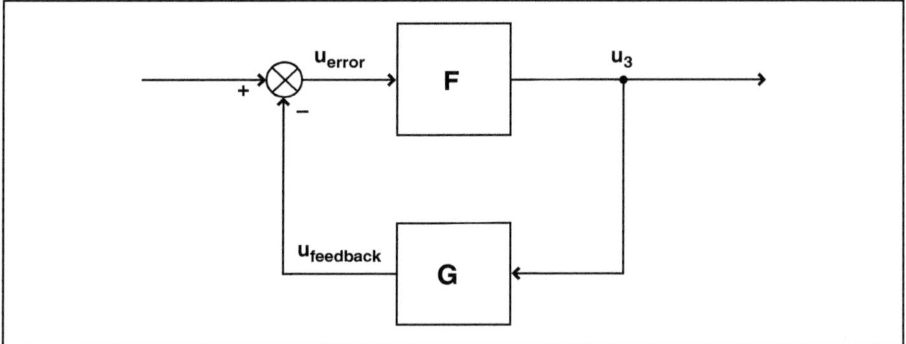

Fig. A2.3. Control system simplified to one forward gain F and a return gain G.

condensed into so-called first order and second order elements and all control engineering is built around them.

Since the loop gains are greater than one, the question of the stability of a loop arises. For this, one cuts the loop open and considers a signal traveling around the loop. The loop gains must be smaller than one or, using the Laplace transformation, the real part must be smaller than minus one. The minus sign comes only because there is another minus sign at the mixing point of Fig A2.3. The former is the essence of the Nyquist stability theorem from which the root locus follows.

Mason's loop rule is important as it determines the overall TF (Transfer Function) from the single gains F and G. In words: the overall TF equals the forward gain divided by one plus the loop gain. As a formula :

$$L_g = F\,G$$

$$T\,F = \frac{F}{1 + L_g} = \frac{F}{1 + F\,G}$$

As mentioned, in Nyquist stability one considers the negative input to the mixing point and therefore one considers the minus one point for stability.

The whole discussion resembles excitation (the instability) of radio transmitters and oscillators where, for oscillation, one always considers two balances, the amplitude balance and the phase balance.

The amplitude balance considers the gain of one, that is the point minus one in the control engineering sense for the onset of oscillation and the phase balance for frequency.

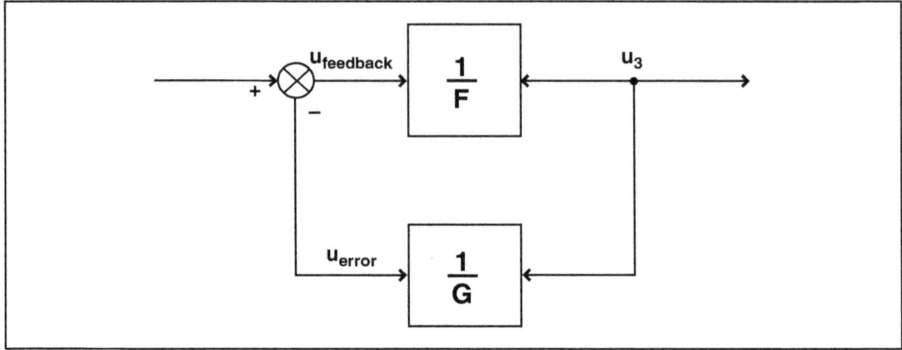

Fig. A2.4. Inverted control system with forward gain 1/G and return gain 1/F.

Control loops are important in control engineering, but they can be turned around. This is shown in Fig A2.4 where the gains become reciprocal, that is 1/F and 1/G. Hence the role of error variable and return variable are interchanged. We call this an inverted control loop and the whole process is the inversion of Block diagrams. All the loop gains become reciprocals by inversion and, in particular, the roles of differentiation and integration are interchanged. So, the all procedure could be used to avoid derivation of a signal.

Note that if the loop gain is large compared to one in the original BD, it will be small in the inverted BD. This process is used for certain tasks in control engineering and it's relation to BG is showed in [GDT 2000]. In that regard, pages 47 to 54 are especially of interest with their discussion of a "causal path".

The overall FT in the inverted case becomes:

$$L_g = \frac{1}{L\,G}$$

$$F\,T = \frac{\frac{1}{G}}{1 + \frac{1}{F\,G}} = \frac{F}{1 + F\,G}$$

So the overall Transfer Function is the same in the direct case (Fig A2.3) and the inverted case (Fig A2.4).

One point at issue is whether loop inversion is really possible. One can say that it may not be possible with real apparatus, but it is certainly possible as a computation scheme.

Appendix 3

Historical Notes

BG (Bondgraphs) were invented in the 1960s at MIT (Cambridge, Mass, USA) by Prof. H. Paynter together with his assistants D. Karnopp and R. Rosenberg. Thoma has used them since 1968 when he was at MIT giving a course on Fluid Power Control and heard Karnopp explain them.

Born of the desire to simulate hydraulic control systems containing mechanical and electrical parts, BG started with electrical and equivalent circuits. In BG, the picture symbols of electronics are replaced by simple letters, such as R = resistor, C = capacitor, I = inductor, and so on. There are also parallel and series junctions. All of this is employed to write an alphanumeric code, a sequence of letters and numbers that could be input into the computers of the era.

Now (2006) we use mostly TWENTESIM (also called 20-sim), which allows BGs to be designed on the computer screen. The program then formulates the equations automatically, solves them and displays simulation curves on the screen. Sometimes it is necessary to give the formulation a little manual assistance, but this is quick and easy.

The supplier of TWENTESIM previously had a program called TUTSIM, which Thoma used extensively in Waterloo, Canada. But this was more difficult to handle since the equations had to be transformed into block diagrams, therefore it is now obsolete.

Here we have reminded the reader only of the fundamental definitions, but for a complete description we recommend [THOMA 1975] or the more recent [THOMA 1990], or [KMR 2000]. For fluid simulation in particular, see [THOMA 2000]. A very interesting recent collection of BG techniques by several authors in french is [GDT 2000].

Thoma started out with hydrostatic transmissions and fluid power control and is now also interested in compressible fluid lines as a corollary to electric wires. Another point of interest is the use of entropy as thermal charge. It provides a much better understanding of thermodynamics and clears up many misconceptions, including some points from biology.

Epilogue

Now that our journey through thermodynamics and systems engineering is over, we pause for a moment's reflection.

We, the authors, have always been impressed by the deep unity of classical physics and engineering. This unity results in a beautiful edifice that includes mechanics, electricity and thermics and allows us to see the world with new eyes. Even chemistry, biology and information systems fit into our picture, which is so aptly expressed by Bond Graphs.

We now say goodbye to the reader and offer a final piece of advice: ne relâchez pas vos efforts car il y a tant de choses à étudier (don't give up your efforts because there are so many things that need to be studied).

The authors

Io son Beatrice, che ti faccio andare;
amor mi mosse, che mi fa parlare.

Dante Alighieri

I am Beatrice, who makes thee go;
love which moved me, makes me speak.

Concepts

Concept	Symbol	Structure	Unit
Advancement of reaction	ξ		
Area	A	$[\mathrm{m}^2]$	
Bond: symbol of transmission of power, product of two factors e and f, time derivative of energy, between two elements E_i,	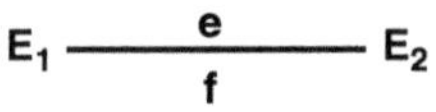 $E_1 \xrightarrow[\ f\ ]{\ e\ } E_2$	$\dot{E} = \mathrm{ef}$	
Causality effort: the effort goes from left to right; the flow from right to left. E_2 takes the effort and computes the flow.	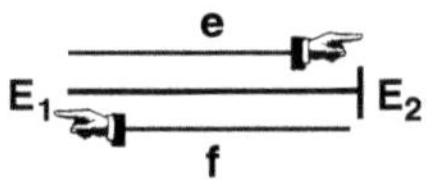 $E_1 \underset{f}{\overset{e}{\longrightarrow}} E_2$	$\dot{E} = \mathrm{ef}$	
Causality flow: the effort goes from right to left; the flow from left to right. E_1 takes the effort and computes the flow.	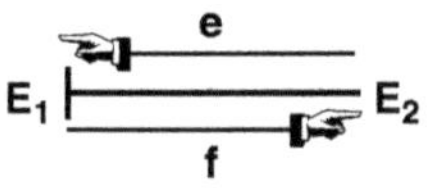 $E_1 \underset{f}{\overset{e}{\longrightarrow}} E_2$	$\dot{E} = \mathrm{ef}$	
Capacitor: generalized capacitor which integrates flow.	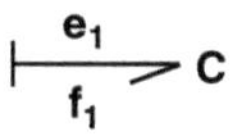 $\xrightarrow[f_1]{e_1} C$	$e = \dfrac{1}{C} \displaystyle\int \mathrm{f}\,dt$	
Capacitor field: multiport energy storage.	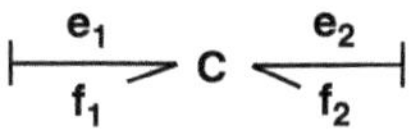 $\xrightarrow[f_1]{e_1} C \xleftarrow[f_2]{e_2}$	$\begin{pmatrix} e_1 \\ e_2 \end{pmatrix} = f \begin{pmatrix} q_1 \\ q_2 \end{pmatrix}$	

Concept	Symbol	Structure	Unit
Chemical potential	μ	$[\mathrm{Jn}^{-1}]$	
Combined IC multiport: multiport energy storage.	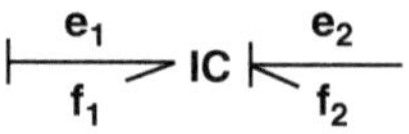	$\begin{pmatrix} e_1 \\ f_2 \end{pmatrix} = f \begin{pmatrix} p_1 \\ q_2 \end{pmatrix}$	
Detector of effort: produces a signal s for a controller.	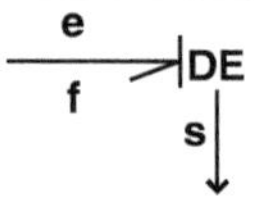	$s = K e$	
Detector of flow: produces a signal s for a controller.	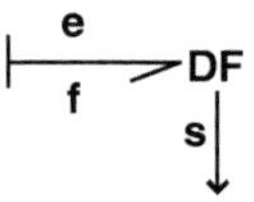	$s = K f$	
Diameter	D	[m]	meter
Displacement	x	[m]	meter
Electric charge	q	[C]	coulomb
Electric current	i	[A]	ampère
Electric voltage	u	[V]	volt
Enthalpy	H	[J]	joule
Enthalpy flow	$\dot{H}$	[W]	watt
Entropy	S	[J]	joule
Entropy density	s	$[\mathrm{Jkg}^{-1}]$	
Entropy flow	$\dot{S}$	[W]	watt
Force	f	[N]	newton
Forward affinity	A_f	$[\mathrm{Jn}^{-1}]$	
Global specific heat at constant pressure	C_p	$[\mathrm{JK}^{-1}]$	
Global specific heat at constant volume	C_v	$[\mathrm{JK}^{-1}]$	

Concept	Symbol	Structure	Unit
Gyrator: transmits the power in the same or in an other energy domain.	e / f →DF, s↓	$f_2 = Ke_1$ $f_1 = Ke_2$	
Heat exchanger:	**HEXA**		
Inductor: integrates effort to give flow.	e / f →I	$f = \dfrac{1}{I} \displaystyle\int e\,dt$	
Inductor field: multiport energy storage.	e_1 / f_1 →I← e_2 / f_2	$\begin{pmatrix} f_1 \\ f_2 \end{pmatrix} = f \begin{pmatrix} p_1 \\ p_2 \end{pmatrix}$	
Internal energy	U	$[J]$	joule
Internal energy flow	$\dot{U}$	$[W]$	watt
Junction effort: receives an effort and distributes it among elements in which flow is equal.	e_1 / f_1 →1, e_2 / f_2, e_3 / f_3	$e_3 = e_1 + e_2$ $f_1 = f_2 = f_3$	
Junction flow: receives a flow and distributes it among elements in which effort is equal.	e_1 / f_1 →1, e_2 / f_2, e_3 / f_3	$f_3 = f_1 + f_2$ $e_1 = e_2 = e_3$	
Length	L	$[m]$	meter
Level	N	$[m]$	meter
Mass	m	$[kg]$	kilogramme
Mass flow	$\dot{m}$	$[kgs^{-1}]$	
Mechanical power	$\dot{E}$	$[W] = [Js^{-1}]$	watt
Memristor: charge or impulse controlled resistor.	e / f →M	$e = f_M(q)f$ $f = f_M(p)e$	

Concept	Symbol	Structure	Unit
Modulated gyrator: gyrator modulated by a signal s from a controller.	$\xrightarrow[f_2]{e_1}$ MGY $\xrightarrow[f_2]{e_2}$ s↑	$f_2 = K(s)e_1$ $f_1 = K(s)e_2$	
Modulated source of effort:	MSE $\xrightarrow[f]{e}$ s↑	$e = K(s)$	
Modulated source of flow: source modulated by a control signal s.	MSF $\xrightarrow[f]{e}$ s↑	$f = K(s)$	
Modulated transformer: transformer modulable by a signal s from a controller.	$\xrightarrow[f_2]{e_1}$ MTF $\xrightarrow[f_2]{e_2}$ s↑	$e_2 = K(s)e_1$ $f_1 = K(s)f_2$	
Molar flow	$\dot{n}$	$[\mathrm{ns}^{-1}]$	
Molar mass	n	$[\mathrm{n}]$	
Non-linear resistor: dissipator or power sink.	$\xrightarrow[f]{e}$ R	$e = \mathbf{R}(f)\mathbf{f}$ $f = \dfrac{1}{\mathbf{R}(f)}e$	
Power	P_u	$[\mathrm{W}]$	watt
Pressure	P	$[\mathrm{Pa}]$	pascal
Resistor: dissipator or power sink.	$\xrightarrow[f]{e}$ R	$e = \mathrm{R}\,f$ $f = \dfrac{1}{\mathrm{R}}e$	
Resistor source: irreversible source of entropy.	$\xrightarrow[f]{e}$ RS $\xrightarrow[\dot{S}]{T}$		
Resistor source field: multi-energy dissipation.	$\xrightarrow[f_1]{e_1}$ RS $\xrightarrow[f_2]{e_2}$ $\dot{S}\mid T$		
Resistor to convection:	**RECO**		

Concept	Symbol	Structure	Unit
Reverse affinity	A_r	$[\text{Jn}^{-1}]$	
Sign: the half arrow gives the conventional positive direction of the transmission of power.	$\xrightarrow[\ f\ ]{\ e\ }$		
Signal: one of the two variables is disregarded so we only get a signal s.	$\xrightarrow{\ s\ }$		
Source of effort: supplies or withdraws power.	SE $\xrightarrow[\ f\ ]{\ e\ }$	$e = \text{constant}$	
Source of flow: supplies or withdraws power.	SE $\xrightarrow[\ f\ ]{\ e\ }$	$f = \text{constant}$	
Specific heat per mass: at constant pressure	C_p	$[\text{Jkg}^{-1}\text{K}^{-1}]$	
Specific heat per mass: at constant volume	C_v	$[\text{Jkg}^{-1}\text{K}^{-1}]$	
Specific enthalpy	h	$[\text{Jkg}^{-1}]$	
Specific volume	ν	$[\text{m}^3\ \text{kg}^{-1}]$	
Speed of reaction	$\dot{\xi}$	$[\text{s}^{-1}]$	
Steam quality	X	$[0\text{-}1]$	
Switch: switches between sources.	**SW**		
Temperature	T	$[\text{K}]$	kelvin
Thermal conductance	K_c	$[\text{WK}^{-1}]$	
Thermal conductivity	λ	$[\text{Wm}^{-1}\text{K}^{-1}]$	

Concept	Symbol	Structure	Unit
Thermal power	$\dot{E}$	$[W] = [Js^{-1}]$	watt
Thermal power by conduction	$\dot{Q}$	$[W]$	
Thermofluid machine	TEFMA		
Thickness	e	$[m]$	meter
Transformer: transmits the power in the same or in another energy domain.	SE $\vdash\!\!\dfrac{e}{f}\!\!\longrightarrow$	$e_2 = Ke_1$ $f_1 = Kf_2$	
Volume	V	$[m^3]$	
Volume flow	$\dot{V}$	$[m^3\ s^{-1}]$	
Volumic mass	ρ	$[kgm^{-3}]$	
Width	l	$[m]$	

Symbols

Symbol	Concept	Structure	Unit
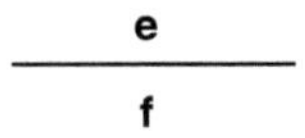 e f	**Bond:** Transmission of power, time derivative of energy, as a product of two factors: effort e and flow f.	$\dot{E} = \mathrm{ef}$	
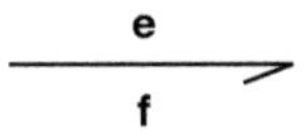 e f	**Signed bond:** The half arrow gives the conventional positive direction of the transmission of power.	$\dot{E} = \mathrm{ef}$	
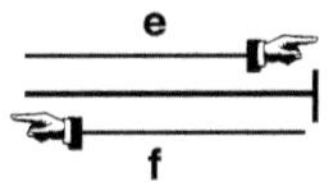 e f	**Effort causality:** The stroke is always at the end where effort is acting.		
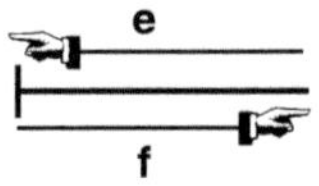 e f	**Flow causality:** The stroke is always at the end where effort is acting.		
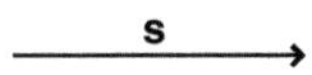 s	**Signal:** One of the two variables is disregarded so we only get a signal s.		

Symbol	Concept	Structure	Unit
e ⊢→ C (f)	**Capacitor:** Generalized capacitor which integrates flow.	$e = \dfrac{1}{C} \displaystyle\int f\, dt$	
e →| I (f)	**Inductor:** Generalized inductor which integrates effort.	$f = \dfrac{1}{I} \displaystyle\int e\, dt$	
e →| R (f)	**Resistor:** Dissipator or power sink.	$e = \mathbf{R}f$ $f = \tfrac{1}{\mathbf{R}}e$	
e →| R (f)	**Non-linear resistor:** Dissipator or power sink.	$e = \mathbf{R}(f)f$ $f = \tfrac{1}{\mathbf{R}(f)}e$	
e →| M (f)	**Memristor:** Charge or impulse controlled resistor.	$e = f_M(q)f$ $f = f_M(p)e$	
e → RS ⊢ T / Ṡ (f)	**Resistor source:** Irreversible source of entropy.		
SE — e →| (f)	**Source of effort:** Supplies or withdraws power.	$e = \text{constant}$	
SF ⊢ e → (f)	**Source of flow:** Supplies or withdraws power.	$f = \text{constant}$	
MSE — e →| (f), s↑	**Modulated source of effort:** Source modulated by a control signal s.	$e = K(s)$	
MSF ⊢ e → (f), s↑	**Modulated source of flow:** Source modulated by a control signal s.	$f = K(s)$	
e →| DE (f), s↓	**Detector of effort:** Produces a signal s for a controller.	$s = Ke$	

Symbol	Concept	Structure	Unit
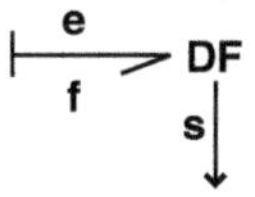 **DF** (e, f, s)	**Detector of flow:** Produces a signal s for a controller.	$s = Kf$	
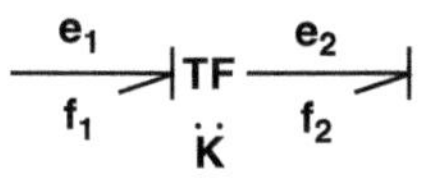 **TF** $\ddot{K}$ (e_1, f_1, e_2, f_2)	**Transformer:** Transmits power in the same or in another energy domain.	$e_2 = Ke_1$ $f_1 = Kf_2$	
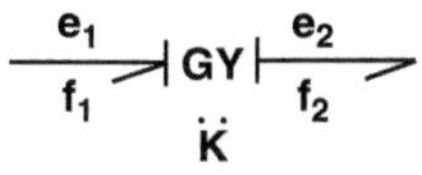 **GY** $\ddot{K}$ (e_1, f_1, e_2, f_2)	**Gyrator:** Transmits power in the same or in another energy domain.	$f_2 = Ke_1$ $f_1 = Ke_2$	
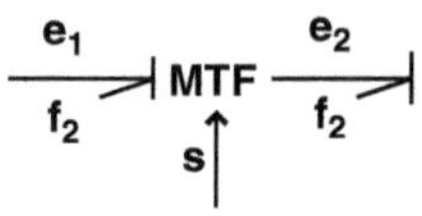 **MTF** s (e_1, f_2, e_2, f_2)	**Modulated transformer:** Transformer modulable by a signal s from a controller.	$e_2 = K(s)e_1$ $f_1 = K(s)f_2$	
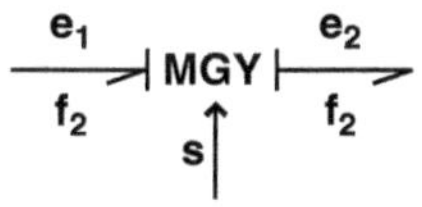 **MGY** s (e_1, f_2, e_2, f_2)	**Modulated gyrator:** Gyrator modulable by a signal s from a controller.	$f_2 = K(s)e_1$ $f_1 = K(s)e_2$	
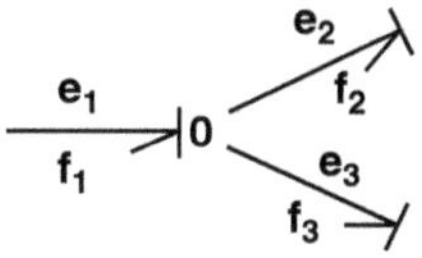 **0** (e_1, f_1, e_2, f_2, e_3, f_3)	**Effort junction:** Receives an effort and distributes it among elements in which flow is equal.	$e_3 = e_1 + e_2$ $f_1 = f_2 = f_3$	
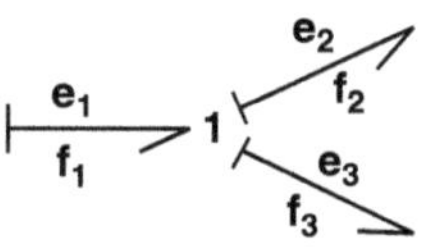 **1** (e_1, f_1, e_2, f_2, e_3, f_3)	**Flow junction:** Receives a flow and distributes it among elements in which effort is equal.	$f_3 = f_1 + f_2$ $e_1 = e_2 = e_3$	
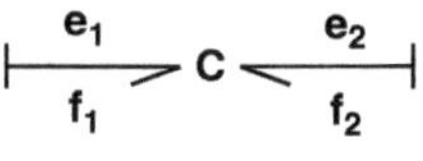 **C** (e_1, f_1, e_2, f_2)	**Multiport capacitance:** Multi-energy storage.	$\begin{pmatrix} e_1 \\ e_2 \end{pmatrix} = f \begin{pmatrix} q_1 \\ q_2 \end{pmatrix}$	
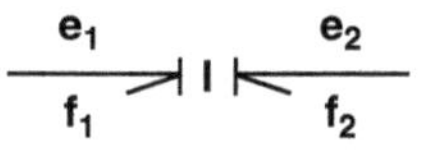 **I** (e_1, f_1, e_2, f_2)	**Multiport inductance:** Multi-energy storage.	$\begin{pmatrix} f_1 \\ f_2 \end{pmatrix} = f \begin{pmatrix} p_1 \\ p_2 \end{pmatrix}$	

Symbol	Concept	Structure	Unit
$\xrightarrow[f_1]{e_1}$ IC $\xleftarrow[f_2]{e_2}$	**Combined IC multiport:** Mixed multi-energy storage.	$\begin{pmatrix} e_1 \\ f_2 \end{pmatrix} = f \begin{pmatrix} p_1 \\ q_2 \end{pmatrix}$	
$\xrightarrow[f_1]{e_1}$ RS $\xleftarrow[f_2]{e_2}$ $\dot{S}\mid T$	**Resistance source:** Multi-energy dissipation.	$T\dot{S} = e_1 f_1 - e_2 f_2$	
$\mathbf{SW}^1$	**Switch:** Switches between sources.		
RECO	**Resistance to convection**		
HEXA	**Heat exchanger**		
TEFMA	**Thermofluid machine**		
A	**Area**	$[\mathrm{m}^2]$	
A_f	**Forward affinity**	$[\mathrm{Jn}^{-1}]$	
A_r	**Reverse affinity**	$[\mathrm{Jn}^{-1}]$	
c_p	**Specific heat per mass** at constant pressure	$[\mathrm{Jkg}^{-1}\mathrm{K}^{-1}]$	
C_p	**Global specific heat** at constant pressure	$[\mathrm{JK}^{-1}]$	
c_v	**Specific heat per mass** at constant volume	$[\mathrm{Jkg}^{-1}\mathrm{K}^{-1}]$	
C_v	**Global specific heat** at constant volume	$[\mathrm{JK}^{-1}]$	

---

[1] Here Thoma breaks his own rule [THOMA 1990]: the symbol SW should have 4 or more letters (i.e. SWIT). Indeed, SW is so important that it should be a BG symbol.

Symbol	Concept	Structure	Unit
ξ	Advancement of reaction		
$\dot{\xi}$	Speed of reaction		
D	Diameter	$[\text{m}]$	meter
$\dot{E}$	Thermal power	$[\text{W}] = [\text{Js}^{-1}]$	watt
$\dot{E}$	Mechanical power	$[\text{W}] = [\text{Js}^{-1}]$	watt
e	Thickness	$[\text{m}]$	meter
f	Force	$[\text{N}]$	newton
H	Total enthalpy	$[\text{J}]$	joule
$\dot{H}$	Enthalpy flow	$[\text{W}]$	watt
h	Specific enthalpy	$[\text{Jkg}^{-1}]$	
i	Electric current	$[\text{A}]$	ampere
K_c	Thermal conductance	$[\text{WK}^{-1}]$	
l	Width	$[\text{m}]$	
L	Length	$[\text{m}]$	
λ	Thermal conductivity	$[\text{Wm}^{-1}\text{K}^{-1}]$	
m	Mass	$[\text{kg}]$	kilogram
$\dot{m}$	Mass flow	$[\text{kgs}^{-1}]$	
μ	Chemical potential	$[\text{Jn}^{-1}]$	
n	Molar mass	$[\text{n}]$	
$\dot{n}$	Molar flow	$[\text{ns}^{-1}]$	
N	Level	$[\text{m}]$	meter
P	Pressure	$[\text{Pa}]$	pascal
P_u	Power	$[\text{W}]$	watt

Symbol	Concept	Structure	Unit
$\dot{Q}$	Thermal power conduction	[W]	
q	Electric charge	[C]	coulomb
ρ	Volumic mass	$[\text{kgm}^{-3}]$	
S	Entropy	[J]	
$\dot{S}$	Entropy flow	[W]	
s	Entropy density	$[\text{Jkg}^{-1}]$	
T	Temperature	[K]	kelvin
u	Electric voltage	[V]	volt
U	Internal energy	[J]	joule
$\dot{U}$	Internal energy flow	[W]	watt
V	Volume	$[\text{m}^3]$	
$\dot{V}$	Volume flow	$[\text{m}^3\,\text{s}^{-1}]$	
ν	Specific volume	$[\text{m}^3\,\text{kg}^{-1}]$	
x	Displacement	[m]	meter
X	Steam quality	[0-1]	

Index